BRIDGESCAPE

Second Edition

THE ART OF DESIGNING BRIDGES

Frederick Gottemoeller

WILEY

John Wiley & Sons, Inc.

Cover Photo: Boston's new Zakim Bridge carries I-93 and an interchange ramp across the Charles River. The Swiss bridge engineer Christian Menn developed the concept and did the preliminary design.

This book is printed on acid-free paper. ∞

For general information on our other products and services or for technical support, please contact our Customer Care Department within the United States at (800) 762-2974, outside the United States at (317) 572-3993 or fax (317) 572-4002.

Wiley also publishes its books in a variety of electronic formats. Some content that appears in print may not be available in electronic books. For more information about Wiley products, visit our web site at www.wiley.com.

Library of Congress Cataloging-in-Publication Data:
Gottemoeller, Frederick.
    Bridgescape: the art of designing bridges/Frederick Gottemoeller.—2nd ed.
        p. cm.
    ISBN 0-471-26773-2 (Cloth)
    1. Bridges—Design and construction.   2. Architecture—Aesthetics.   I. Title.
TG300.G67   2004
725'.98—dc222
                                                                2003017779

Printed in the United States of America

10 9 8 7 6 5 4 3 2 1

Contents

FOREWORD TO THE FIRST EDITION

In the United States of the early 1950s, anyone beginning to practice structural engineering faced a society in which new construction was underway at an unprecedented rate; ironically, however, education in engineering design was almost nonexistent. In particular, once the interstate system began to be built, bridges also began to appear in vast numbers with, sadly, very few attracting attention as anything more than "mere" utilitarian objects. Education in structural analysis and the dimensioning of structural elements took precedence over design—design, that is, in the sense of setting the form and paying attention to appearance. There was almost no literature on bridge design in the United States; post–World War II structural engineers had little knowledge of the great structures of the recent past.

More recently, many articles on bridge aesthetics have appeared; but in reality these works, especially those by engineers, are on design. The most complete, annotated bibliography on bridge aesthetics illustrates well the absence of bridge design writings right after World War II. Of the 244 entries, only 65 were written in the 100 years before 1950, and only 8 were written in the 1950s. By contrast, 34 appeared in the 1960s, 69 in the 1970s, and 68 between 1980 and 1990. What all of this writing needed was some work that would organize and focus these diverse ideas into a book that could be used directly by both educators and practicing engineers. Frederick Gottemoeller has done just that in this carefully illustrated volume.

Three features mark this book:

First, he has organized design around two sets each of five criteria, one primarily structural and having major influences on appearance, and the other only secondarily structural and having less influence. Moreover, by putting each of the ten criteria in order of importance, he focuses our attention on priorities and controlling ideas rather than on merely a long checklist. At the same time, he deals with a wide variety of details but in an ordered way.

Second, he has illustrated all of the ideas with clear photos of actual bridges, as well as with elegant drawings. The illustrations are not mere decoration, as they are in so many engineering texts, but are linked carefully to the text. As such they not only make clear Mr. Gottemoeller's ideas, but overall they fit his central theme that all parts of the structure are to be at once both useful and attractive.

Third, he continually emphasizes the idea that bridge appearance is the province of the structural engineer and that the primary features of bridges are structural. It is just those structural ideas that can best lead to more pleasing bridges. The profession is called upon by society to design bridges that will please not only the present population but also its posterity. The book thus expands the vision of engineering to include design in its most general sense, which implies, as he makes clear, that engineers become articulate spokespersons for the profession both in their handling of the materials for design and in their interactions with the general public.

I have worked with Fred Gottemoeller on bridge design issues for over 10 years and I know that our profession will welcome his book. It is sorely needed. Together, he and I have collaborated with many forward-looking state bridge engineers—all of whom are wrestling with the problem of responsible design: bridges that will appeal visually to the public while at the same time satisfying the stringent requirements of performance and cost. Bridges are one major example of that central cultural issue in an open society of public accountability.

That feature of accountability represents another aspect of Fred Gottemoeller's talents reflected in this book. He is deeply concerned with the problem of involving local people in the design process such that they can better understand engineering, and engineers can better appreciate the need for striking appearance in bridges. Mr. Gottemoeller devotes much of his professional practice to interactions with the public and this book lays a strong groundwork for such activity. The book, being free from the dense jargon of engineering while being beautifully illustrated, will provide common ground for citizen–engineer mutual education.

The author's education in both engineering and architecture allows him to introduce ideas common to one profession that stresses aesthetics and from the other profession that centers upon technique. When these two sets of ideas merge, as they do in this book, a new synthesis emerges that opens up opportunities for structural engineers. The general public is aware of the great suspension bridges; this book should make them look again at what they now see as the more mundane works of bridge engineering.

David P. Billington
Princeton University

Foreword to the Second Edition

In recent years myriad books titled "Bridges" have been thrown on the market, usually cursory collections of brilliant photographs of the same visual highlights of the same landmark bridges. Though these books do bring bridges to the attention of the broad public, they simultaneously insinuate that creating a bridge is the result of a stroke of genius accessible only to a very few inspired artists or architects, and that only landmark bridges are worth the effort and the attention of such gifted designers.

Fred Gottemoeller's book tackles what bridge design is really about. He addresses all bridges that have an impact on our natural and cultural environment, not just the landmark bridges, and shows how to arrive at solutions that are technically, economically, and aesthetically appropriate.

Bridges are prominent and lasting. They can enrich or spoil their environments for a long time. That includes the small and medium-size "workhorses" (as the author calls them) of our transportation system for rail and road that, by their sheer number, have a major impact on the environment. These countless overcrossings are usually so dull and mean, each of them a missed opportunity to expose millions of travellers to creative bridge design. Though other products of human creativity, such as cars, furniture, or even single-family houses are, when compared with bridges, only ephemeral, each has its own visually adept profession shaping its design. Bridges do not. Writing about and teaching bridge design at our universities is still limited to analysis and dimensioning, skipping the conceptual design that determines form and that must precede analysis and predetermine its output.

My identification of this gap should by no means be misinterpreted as a call for visual bridge designers to embellish and decorate what some mere engineer has left behind, nor for bridge design competitions amongst architects.

No, holistic and high-quality bridges only can result from a profound knowledge in structural engineering, from materials through analysis to manufacture combined in one-and-the-same person with creativity, imagination, and curiosity. The shape of a bridge must result from its purpose, its site, its materials, its flow of forces, and its erection process, so that each bridge may be considered some day as a contribution to its culture. Those who skip the conceptual design phase of a bridge by recycling a standard design miss the best and most stimulating part of structural engineering. Inspecting the site, preparing and rejecting hundreds of sketches, trying and improving again and again, and finally some years later inspecting the outcome provides the engaged bridge designer with that indescribable feeling God must have felt after creating the earth.

If all educated engineers possess that knowledge needed to conceive a bridge, and if most human beings by birth are gifted with intuition and creativity, why then are there are so many mean and so few good bridges? Fred Gottemoeller must be of the opinion—and I fully agree with him—that being able to combine knowledge and creativity in conceptual design must be learned and trained, just as with the art of composing music. Music composition must integrate the knowledge of playing an instrument with artistic inspiration, so that finally rhythm and melody merge, as should function and form in a bridge.

The contribution that Fred Gottemoeller makes by unselfishly concentrating his attention on small- and medium-sized "everyday" bridges cannot be overestimated. Writing on spectacular large bridges is much simpler and more attractive because a small bridge leaves much more room for subjective choice than a large one. He visualizes and describes these multiple choices and leads you step-by-step in the right direction by opening your eyes. One can understand only what one sees!

All those eager to learn and thereafter to practise creative bridge design and all those eager to demonstrate that a stimulating bridge needs neither to cost more nor be less durable than a standard type will welcome this book. Particularly valuable is its call for more engineering design competitions. The designs of museums, administrative buildings, and even schools are determined through competitions, but that of infrastructure, especially bridges, is too often determined by least price. A sensible and lightweight bridge may cost more than its bulky standard counterpart, but this difference in cost must result from less material consumption and correspondingly more labor, which makes it ecologically and socially more valuable as well as more beautiful.

May this book make structural engineers aware of the unique opportunities of their profession and evoke bridges that are assets to their environments, societies, and cultures.

Jörg Schlaich
Stuttgart

PREFACE

In 39 years of working in, and managing engineering organizations, I have found that most engineers recognize the importance of good appearance. Most would like to do a better job with the aesthetics of their bridges. They are just not sure how.

"Aesthetics" is a mysterious subject to most engineers, not lending itself to the engineer's usual tools of analysis, and rarely taught in engineering schools. Being both an architect and engineer, I know that it is possible to demystify aesthetics in the minds of engineers. The works of engineers like Robert Maillart, Christian Menn, Jean Muller, Jorg Schlaich and others proves that engineers can understand aesthetics. Unfortunately, such examples are too rare. The principles of bridge aesthetics should be made accessible to all engineers.

That was my goal for the first edition of this book. Engineers' reception of that book confirmed the value of that goal. The second edition retains the same goal, and seeks to make the book even more directly useful to practicing engineers and engineering students by:

- Providing more discussion on the "why" behind the various recommendations and using more case studies of actual bridges;

- Integrating the developing procedures of Context-Sensitive Design/Context-Sensitive Solutions into the suggested design process;

- Adding a section on moveable bridges;

- Adding information on the contribution of aesthetics to environmental impact studies;

- Expanding and updating the discussion of community participation, a field that continues to grow as the public demands a greater role in bridge design;

- Updating the illustrations;

- Adding the insights gained by seven years of practice in the field since the first edition was written.

Bridge design is an art, an art that uses science and mathematics to support many of its judgments. Other choices are made during the design process that science cannot help, such as decisions about appearance. This book will give bridge designers a basis to support those aesthetic judgments, so that their decisions about appearance can be just as definitive as the judgments they make about structural members, safety, or cost. This knowledge can then be a tool for designers to use when justifying their recommendations to managers and contractors.

Aesthetic ideas, like any other, change over time as people working in the field gain new insights, respond to new materials and technologies, and learn from their own experiences and those of others. This book will prepare engineers to respond to these developments and contribute aesthetic insights of their own. The book's strategy is to provide engineers with basic principles and tools of analysis that will help them think through questions of appearance. The case studies then show how those ideas have been applied to example bridges.

Appearance is a subject that does not benefit from hard-and-fast rules. Each bridge is unique, and should be treated that way. This book does more than present guidelines, rules of thumb, and comparative examples; it also encourages engineers to make their own judgments about what looks good and what does not, and which guidelines apply to a particular structure and which do not.

Engineers should take away from this book a permanent commitment to consider each structure's appearance, starting with the structure's "bones": the girders, piers, and abutments. The aesthetic impact of a bridge is primarily a product of the structural members themselves. Details and color are important, but secondary.

Every year, thousands of engineers design thousands of small and medium-sized bridges. These bridges are the "workhorses" of our transportation systems. Because there are so many of them, they dominate the appearance of most of our highways, neighborhoods, towns, and landscapes. Large bridges, because of their size and prominence, often receive more attention; however, smaller bridges, taken together, have the greater total impact. This book focuses on the workhorse bridges—bridges with spans less than 500 feet.

Most of the examples shown, and some of the discussion on perception, involve highway bridges and highway conditions. The principles described, however, are equally applicable to transit bridges. The sections on viaducts and ramps, for example, have a clear application to elevated transit structures.

Although this book is aimed at engineers and engineering students, other professionals, such as architects, landscape architects, artists, and members of the general public concerned about the appearance of bridges, will find it a useful introduction to the aesthetic aspects of bridge engineering.

ACKNOWLEDGMENTS

This book grew out of observations of bridges in the United States and abroad beginning when I was a student of Architecture at Carnegie Mellon University. Max Bill's book *Robert Maillart: Bridges and Construction* (Praeger Publishers, 1969 ed.) had kindled my interest in bridges, and led to the addition of Civil Engineering to my studies at Carnegie Mellon.

I found an intellectual framework for my observations in the work of David P. Billington of Princeton University, particularly in his book *The Tower and the Bridge* (Basic Books, 1983) and his works on Maillart—*Robert Maillart's Bridges: The Art of Engineering* (Princeton University Press, 1979) and *Robert Maillart and the Art of Reinforced Concrete* (MIT Press, 1989). David has since become a source of valuable support, a collaborator on various seminar projects, and a friend.

The bridges of the German engineer Jorg Schlaich, and his thoughts on conceptual engineering, are continuing sources of inspiration. Several of his pedestrian bridges and one of his railroad bridges are included in Chapter Six. Another German engineer, Fritz Leonhardt, through his book *Brucken* (The MIT Press, 1982),[4] has also had a major influence on me. I am particularly indebted for his willingness to allow me the use of several illustrations from his book.

Since the publication of the first edition of this book, several masterworks by the Swiss bridge engineer Christian Menn have been completed. The Leonard P. Zakim Bridge in Boston appears on the cover of the second edition, and the Sunniberg Bridge near Klosters in Switzerland is the subject of a new case study added to the book. In my opinion, these bridges epitomize the engineering aesthetic to which bridge designers should aspire. Through David Billington's efforts, these works were recognized in an exhibition at the Princeton University Museum of Art in the spring of 2003. The book associated with this show, *The Art of Structural Design* (Princeton University Art Museum, 2003), is an invaluable reference.

Since 1997, I have partnered with architect and urban designer Miguel Rosales in Rosales Gottemoeller & Associates, Inc., a consulting firm specializing in improving the aesthetic quality of transportation facilities. This experience has brought the opportunity to test my ideas on bridges of many sizes in many different kinds of communities.

The original catalyst for the book was work for the Maryland State Highway Administration (SHA). In 1987, Hal Kassoff, then Highway Administrator; the Deputy Chief Engineer for Bridges, Earle "Jock" Freedman; and the Director of the Governor's Office of Art and Culture, Jodi Albright, began a collaborative effort to improve the appearance of Maryland's bridges. David and I were engaged to assist with this effort. The effort produced an international conference on bridge aesthetics, a design competition for the new U.S. Naval Academy Bridge across the Severn River in Annapolis, Maryland, now completed (See Figure 1-22, page 19), a seminar series on bridge aesthetics for Maryland's bridge designers, and *Aesthetic Bridges Users Guide,* SHA's aesthetic design guidelines. I am much indebted to Hal and Jock for permission to use material from the Maryland guidelines in preparing the first edition of this book.

Several other state and provincial bridge engineers have provided information, ideas, and inspiration. I include in this group Ed Wasserman of Tennessee; Roger Dorton of Ontario; Don Flemming and Dan Dorgan of Minnesota; Bob Healy of Maryland, John Smith and Bill Rogers of North Carolina; Stewart Gloyd of Washington, Dan Davis of Arizona, and Jim Roberts of California. A number of consulting engineers have been similar resources, including Thomas D. Jenkins, Alex Whitney and Richard Beaupre of URS Corporation; Charles Diver of Diver Brothers; Conrad Bridges, Art Hedgren, and Theun Van de Veen of HDR International, and Nick Altobrando of Hardesty and Hanover. Working with Don Hildebrandt of LDR International (Columbia, Maryland) on the Woodrow Wilson Bridge in Washington, D.C., has made me more aware of the role urban designers and landscape architects can play in assisting bridge design by helping to define the larger environment.

Marty Burke of Burgess and Niple (Columbus, Ohio), formerly Chair of the Transportation Research Board's (TRB) Subcommittee on Bridge Aesthetics, has been the source of inspiration, information, and valuable criticism. His example of indefatigable effort in the preparation of *Bridge Aesthetics Around the World* by the TRB has kept me going in the face of the usual distractions.

For the first edition Alicia Buchwalter, Mary Kay Chadrue, Cathy Levay, Charlotte Miller, and Sharyn Smith were a great help in preparing the manuscript and researching and organizing the illustrations. For the second edition, the efforts of my assistant and daughter, Hillary Huff, in revising and editing the manuscript and organizing the illustrations were indispensable.

My good friends Jim Truby and Karl Sattler have been invaluable sources of support. Without their encouragement it is unlikely that I would have undertaken the original effort. The strongest acknowledgment must go to my wife Pat. Her patience with stops for photographs, and tolerance of long nights and weekends spent at my trusty Macintosh, and most of all her appreciation and enthusiasm, made this book possible.

Frederick Gottemoeller
Columbia, Maryland

chapter **one**

INTRODUCTION

"It so happens that the work which is likely to be our most durable monument, and to convey some knowledge of us to the most remote posterity, is a work of bare utility; not a shrine, not a fortress, not a palace but a bridge."[1]

—MONTGOMERY SCHUYLER, 1883, WRITING ABOUT JOHN ROEBLING'S BROOKLYN BRIDGE

THE AESTHETIC DIMENSION OF BRIDGE ENGINEERING

Bridges speak to us.

They speak to us about the places they are or the places they take us. They speak to us about travel: the new wonders to be seen, the money to be made, the time saved, the excitement of the crossing. They speak to us about the skill of their designers and the courage of their builders. Above all, they speak to us about the values and aspirations of the communities, organizations, and persons who build them.

Several years ago, United Airlines wanted to make a television commercial that conveyed the message that its services reached consumers coast to coast. Which visual images did it choose to convey that message? The Brooklyn Bridge (Figure 1-1), to represent the East Coast, and the Golden Gate Bridge (Figure 1-2), to represent the West Coast.

That is not surprising. Stretching across the Mississippi River or Tampa Bay, bridges can become symbols for whole regions. The recent and widespread familiarity of Tampa's Sunshine Skyway (Figure 1-3) shows how quickly an attractive bridge can come to symbolize a region. The old Skyway was just as functional, but it never caught people's imagination. The new Skyway, with its golden cables gleaming in the sun, has become a nationally recognized symbol of the Tampa area.

The Skyway has come to symbolize something else as well. It represents the best and most modern of technology. National advertisers for luxury cars and computers borrow its reputation and try to apply it to their own products. That's not an unusual role for bridges. The Brooklyn Bridge was cited by many in its day as the best of modern technology.

Bridges have spoken different messages throughout history. To the Romans, bridges were psychological as well as physical tools used to extend the emperor's

FIGURE 1-1 *The Brooklyn Bridge, a symbol of New York City and the East Coast (see color insert).*

control (Figure 1-4). Their message was, "I, the Emperor Trajan, by the power of Rome, have built this massive bridge; realize the impossibility of revolt." In medieval times, specialized orders of monks built bridges for the benefit of the community and the greater glory of God (Figure 1-5). Their message was, "We, Les Freres du Pont, by the grace of our Lord, have built this bridge for you; join us in our pilgrimages to His holy shrines." Throughout time, rulers have used bridges to control travel and commerce (Figure 1-6). Their bridges said, "I, the Duke of Cahors, have built this bridge at the edge of my realm; none may pass unless I consent; all who pass must pay my toll."

The thousands of "everyday" bridges convey messages, too. Take the example of the Capital Beltway (I-95/495) around Washington, D.C.. There must be at least 200 bridges on the Capital Beltway, but only two, the Woodrow Wilson and the Ameri-

FIGURE 1-2 *The Golden Gate Bridge, a symbol of San Francisco and the West Coast (see color insert).*

FIGURE 1-3 *Sunshine Skyway, a symbol of Tampa, Florida, and of modern technology (see color insert).*

FIGURE 1-4 *Puente de Alcantara, Spain, circa 100 AD, a symbol of imperial power, was built by Caius Julius Lancer, one of the few Roman engineers of whom we have a record. He is buried near one end of the bridge. His epitaph reads: "I leave a bridge forever to the generations of the world."*

FIGURE 1-5 *Pont d'Avignon, France, 1178–1187, a symbol of Divine charity. By legend, the engineer was a shepherd who was guided by a divine vision. He was canonized as St. Benezet, and was buried in the chapel on the bridge.*

FIGURE 1-6 *Pont Valentre, France, circa 1337, a symbol of military control.*

can Legion bridges over the Potomac River, can be considered large bridges. The other 198, taken together, cost more to build and maintain than these two. The other 198 are seen everyday by hundreds of thousands of people. They are a more prominent part of the daily lives of these people than any of the world-famous monuments of the capital city. Collectively, they have a huge impact on the public's perception of their city.

Unfortunately, most everyday bridges convey a message of apathy and mediocrity (Figure 1-7). Carrying traffic but lacking grace, they are merely functional. They could be much more. Bridges have the ability to arouse emotions—wonder, awe, surprise, or sheer enjoyment of form or color. They could be works of civic art, which would enliven each day's travels and make everyone's journey more pleasant. The Gunnison Road bridge over I-70 west of Denver (Figure 1-8), frames the westward traveler's first view of the Continental Divide, and magnifies his or her awe. The American Society of Civil Engineers awarded this bridge its outstanding Achievement Award in 1972. Today, in the midst of all of the surrounding scenic magnificence, there is an overlook placed southeast of the bridge for the sole purpose of giving travellers a place from which to view the bridge.

Today, in most countries, freedom of travel is an accepted right, and bridges are no longer asked to carry the weight (literally) of government propaganda. Bridges are public works, authorized by the voters and paid for by the tax dollars of the entire community. In this democratic age, when bridges are built by cities and states made up of voting taxpayers, everyone in a community feels the pride involved in the accomplishment. The unspoken thought is: "Isn't it great that we were able to assemble the knowledge, money, and, skill to build this wonderful bridge? And now we are that much closer to our jobs, our homes, our recreation."

People today first ask that their bridges take advantage of modern technology, make wise use of resources, achieve economic efficiency, and be responsible to the environment. Then they go a step further. They want their cities and towns to be attractive places to live. They know the truth of what Mark Twain once said, "We take stock of a city like we take stock of a man. The clothes and appearance are the

FIGURE 1-7 *A message of apathy and mediocrity, arousing only boredom.*

FIGURE 1-8 *A message of pride and skill, the Gunnison Road Bridge over I-70 west of Denver frames the westward traveler's first view of the Rocky Mountains. (see color insert).*

externals by which we judge." They want their bridges to be a positive feature of their cities. In short, they want their bridges to be beautiful.

Indeed, they are willing to act personally on their desires. On the one hand, they have seen too many ordinary or even ugly bridges built in their communities. On the other hand, they have seen bridges like the Sunshine Skyway in their travels, and wonder why they cannot have a bridge of equal quality in their hometown. They are no longer willing to leave bridge design to the professionals. Many citizens today are demanding a voice in deciding specifically where their bridges are built and what their bridges look like.

Public interest in the appearance of bridges is likely to continue to increase. In 2001, the U.S. Federal Highway Administration published a survey[2] showing that "citizens want highway projects that are more sensitive to local communities, including such items as visual appeal." Transportation agencies have learned that public satisfaction with a project's appearance goes a long way toward getting it approved. A virtuous cycle is beginning. The first attractive bridges have increased the demand for still more attractive bridges, which will benefit both the quality of people's lives and the skills and reputations of the professionals involved.

FIVE FUNDAMENTAL IDEAS

This book is based on five fundamental ideas.

All Bridges Make an Aesthetic Impact

When an engineer builds a bridge, he or she creates a visible object in the environment. People see it, and they react to what they see. The bridge will make an impression: of excitement, appreciation, repulsion, or perhaps boredom. Whether or not the engineer has thought about this visual impact, the bridge will have an impact. We call a bridge's power to move and inform its *aesthetic quality*.

People Can Agree on What Is Beautiful for Bridges as Well as for Paintings and Concertos

We often hear that "beauty is in the eye of the beholder." If we take this phrase to mean that our emotional reaction to a work of art is personal and individual, then that is certainly true. However, if we take it to mean that two people cannot agree on what is a good piece of art, then it is not true. Two people may have individual reactions to the Mona Lisa, but they and millions of others can agree that the Mona Lisa is a great painting. Similarly, people can agree on which bridges are more attractive than others.

Figures 1-9 to 1-11 show three pairs of bridges that are similar in site requirements but very different in appearance. Study them for a moment and form your own opinion about which bridge of each pair is the more attractive. Most of us will agree that the second bridge in each pair is the more attractive.

Do the better-looking bridges have characteristics in common? Yes.

- They are simpler, by which is meant there are fewer individual elements; it also means that the elements that are similar in function (such as girders) are similar

FIGURE 1-9.1 *Highway overcrossing 1.*

FIGURE 1-9.2 *Highway overcrossing 2.*

in size and shape. For example, the bridge in Figure 10.1 has three different types of pedestrian fence/railing; the bridge in Figure 10.2 has one.

- The girders are relatively thinner (the ratio of depth to span is smaller).

- The lines of the structure are continuous, which usually means the spans are continuous or appear to be continuous.

FIGURE 1-10.1 *Pedestrian bridge 1.*

FIGURE 1-10.2 *Pedestrian bridge 2.*

FIGURE 1-11.1 *River crossing 1.*

FIGURE 1-11.2 *River crossing 2.*

- The shapes of the structural members reflect the forces on them. They are thickest where the forces are the greatest, and thinner elsewhere, like the bridge in Figure 1-11.2. By contrast it is impossible to understand the complicated structural system of the bridge in Figure 1-11.1.

We will discuss these characteristics further in Chapter Two and show how they can be used to develop new designs.

The ability to reach agreement on the characteristics of better-looking bridges is critical in a democratic society. It is only with some degree of agreement that consensus can be reached for the expenditure of public funds. That is not to say that the appearance of a bridge must be put to a popular vote. It does mean that, through processes of public debate, written criticism of the type prevalent in the other arts, and the example of works by outstanding practitioners, consensus can emerge about what makes a good-looking bridge. Engineers, public works directors, and the public at large can then use this consensus to guide their decisions about the bridges they build.

Engineers Must Take Responsibility for the Aesthetic Impact of Their Bridges

Engineering is the profession in our society given responsibility for designing bridges. Engineers are used to dealing with issues of performance, efficiency, and cost, but they must also be prepared to deal with the issues of appearance as well: symbolism, appropriateness to the site, and beauty. In describing the design of the Benjamin Franklin Bridge over the Delaware River, Henry Petroski writes:

> Before . . . calculations could be made, however, the overall proportions of the structure had to be established. There were no rigid formulas for doing this; it involved a combination of working with the constraints of bridge location and local geology and of making aesthetic judgments about what just looked right. In taking this all-important step of setting the proportions of a major structure, engineers act more like artists than scientists. And only after the overall defining geometry is set down can the theories and formulas of engineering science be applied to the details. (Ralph) Modjeski and (Leon) Moisseiff (the design engineers for the bridge) were each masters of both the art and the science.[3]

The appearance of the bridge is dominated by the shapes and sizes of the elements controlled by the engineer—the structural elements themselves, not by details, color, or surfaces. The Golden Gate Bridge owes its appeal to the graceful shape of its towers and cables, not to its reddish color. If the towers were ugly, painting them red would not make them attractive.

Periodically the suggestion is made that that engineers should delegate responsibility for the appearance of bridges to architects or other visually trained professionals such as urban designers, landscape architects, or sculptors, the notion being that this would result in better-looking structures. This is rarely successful. The usual result is an overly elaborate and overly costly structure, without a significant improvement in appearance. There are two reasons for this. The first is that the visual professional doesn't understand the special limitations and problems of bridges or, even more important, the aesthetic opportunities of modern engineering materials and structural types. So, the designer falls back on the familiar conventions of architecture or sculpture and makes something that looks like a building or a piece of sculpture, not a bridge.

The second reason is that the delegation can never be complete. In the United States, engineers have the professional and legal responsibility for bridges. The organizations that commission, finance, and build bridges are usually run by engineers. None of these people are willing or legally able to delegate complete responsibility for a bridge to a nonengineer, no matter how talented. This usually means that the "important" (structural) aspects of the bridge are reserved for the engineers, while the "window dressing" (surface details, color) is left to the visual professional. Although the details are an important part of a structure's aesthetic impact, it is the structure itself—its spans, proportions, and major elements—that has the largest role in creating its effect. If the engineer does a poor job with these major elements, no amount of architectural add-ons will compensate.

Engineers must control the structural elements—that is their job. However, for the bridge to be a complete aesthetic success, the details, colors, and surfaces have to be in tune with the structural concept. Therefore, the engineers should control those as well.

That said, visual professionals who are willing to take the time to understand the special nature of bridges can offer a positive contribution to engineers by acting as aesthetic advisors or critics. Their role is comparable to that of other specialists, such as geologists or hydrologists, on whom the engineer may call for advice. See Chapter Three for productive ways to involve visual professionals.

Engineers have accepted a responsibility to society for bridge design. For that reason, no engineer would knowingly build a bridge that is unsafe. For the same reason, no engineer should knowingly build a bridge that is ugly.

Aesthetic Ability Is a Skill That Can Be Acquired and Developed by Engineers, as Well as Anyone Else

Aesthetic ability is not some mysterious quality bestowed by fate on a fortunate few. Though many engineers are not well prepared by their education or experience for the visual aspects of their responsibilities, they can learn what makes bridges attractive and they can develop their abilities to make their own bridges attractive.

The outstanding bridges of John Roebling, Gustave Eiffel, and Robert Maillart show that engineers can produce beautiful bridges. In our generation we have the examples of Christian Menn, Jean Muller, Jörg Schlaich, and others. That said, producing beautiful bridges requires the commitment to do so and the willingness to learn how.

Many tools are available to help engineers improve their abilities in the aesthetic aspect of engineering. This book presents a series of guidelines and the principles that underlie them. It is a useful base from which to consider the appearance of structures. But the guidelines are only a beginning. Engineers should form their own opinions about what looks good, and work to develop their own aesthetic abilities. Chapter Two explains how.

Engineers Should Consider Good Appearance Co-equal with Strength, Safety, and Cost

Bridges must safely carry their loads for a long enough time to repay the investment made in them. This basic fact imposes a hierarchy on design decision making:

- Performance: structural capacity, safety, durability, and maintainability

- Cost: construction and maintenance

- Appearance

That said, this statement often leads to the assumption that the three criteria are necessarily in conflict, that one must be sacrificed for the other. This is not true. For example, many people think improved appearance will automatically add cost. They believe good appearance derives from add-ons, like an unusual color, special materials such as stone or brick, or ornamental features. In fact, the greatest aesthetic impact is made by the structural members themselves—the cables, girders, and piers. These things have to be there anyway. If they are well shaped, the bridge will be attractive, without necessarily adding cost.

Many of the greatest bridge engineers, men such as Thomas Telford, Gustave Eiffel, John Roebling, and Robert Maillart, were able to create bridges that are beautiful, structurally sound, and less expensive than any alternative, all at the same time.

In decisions about appearance, as in decisions about strength and durability, the challenge is to achieve an improvement without an increase in cost. It can be done. The Departments of Transportation of Arizona, California, Maryland, New Mexico, Oregon, Tennessee, and Washington (Figure 1-12); the Ontario Ministry of Transport; and others do it routinely.

FIGURE 1-12 *Routine excellence on Virginia's I-66 west of Washington, D.C.*

FIGURE 1-13 *New arch overpasses on the U.S. 59 Houston Gateway in Texas.*

Too often, however, some restriction in the "hard" criteria—for example, a tight budget or an unusual span requirement—is used as an excuse to ignore appearance. The better approach is to consider the restriction a challenge and use it to find the inherent beauty contained within the problem. That is what the engineers of the Texas Department of Transportation did for the bridges in Figure 1-13, taking an almost impossible span-and-clearance problem and turning it into bridges of uncommon grace. When the US 59 Houston Gateway was widened there was no longer room for median piers, the vertical clearance had to be increased, and the frontage roads could not be raised nor could the freeway be lowered. The engineers resolved these contradictory requirements with a quartet of through arches that span 218 feet with a floor system just 12 inches deep. The bridges could have been aesthetic disasters or just more mediocre missed opportunities, like so many urban freeway bridges. Instead, they are objects of civic pride.

Engineers can design bridges that achieve excellence in all three disciplines: performance, cost, and aesthetic quality. The key is to put all three issues on the table at the same time and work on improving them all together, to not sacrifice one for the other. The goal is to bring forth elegance from utility. The product, as engineer and educator David P. Billington has recognized, is structural art[4]. Structural art is an art form equal to architecture and sculpture, but based on the disciplines of engineering.

Integrating the three issues is not always easy. The following sections discuss in more detail the integration of aesthetic quality and structural performance and the integration of aesthetic quality and cost.

AESTHETIC QUALITY AND STRUCTURAL PERFORMANCE

Structural safety cannot be compromised for the sake of appearance. That should go without saying. With that criterion as a base, it is possible to discuss the relationship of performance and appearance.

FIGURE 1-14 *Stauffacher Bridge over the Sihl River in Zurich, Switzerland, 1899, Maillart's starting point for his three-hinged arches.*

Many of the significant advances in bridge appearance have been made by engineers who were pioneering new structural materials, new construction techniques, and new methods of analysis in order to improve bridge performance. The bridges of John Roebling, Gustave Eiffel, and, in our century, Robert Maillart, Christian Menn, and Jean Muller are all examples. The fact that the appearance of their bridges improved at the same time shows that these two qualities have factors in common and that improvements in one aspect can and should improve the other. Discovering improvements that do both is the engineer's challenge.

Robert Maillart's development of the three-hinged arch is a classic example of how this process can work.[5] Maillart began his career of concrete bridge design with the Stauffacher Bridge over the Sihl River in Zurich (Figure 1-14). Designed in the tradition of historic stone bridges, it gives no indication that the stone is merely a facing and that the loads are carried by a hidden concrete barrel arch.

FIGURE 1-15 *Zuoz Bridge over the Inn River, Zuoz, Switzerland, 1901: Maillart's first step.*

His next assignment, in the rural community of Zuoz, was far from the urban sophistication of Zurich. Maillart was faced with fulfilling a request to design a bridge that the tiny community could afford. He decided to do away with the stone covering and develop a shape that would make the most efficient use of the concrete material. The result was a three-hinged arch using a box-shaped arch, the first application of the box shape to concrete construction (Figure 1-15). The structure not only fit the town's budget, it was an instant success with the community, and quickly became a significant landmark.

Unfortunately, after two years, the bridge developed vertical cracks in the spandrel walls, and Maillart was called back to investigate. He determined that the cracks were caused by differential shrinkage between the arch rib and the deck. He also knew that the stresses in the walls due to loading were very small, and he concluded that the cracks posed no danger.

When he was next asked to design a bridge, he wanted to use the same design that had been successful at Zuoz, but he was faced with the challenge of avoiding future cracking. Many engineers would have added material and reinforcing to prevent the cracks. But, as just noted, Maillart had determined that the stresses in the walls were very small; he believed that the bridge could be redesigned without the walls, so he eliminated them. With no walls, there would be no cracks! The result is the bridge over the Rhine River at Tavenasa, which became the prototype for all of his later three-hinged structures (Figure 1-16). From this point forward, he continuously refined the shape of this bridge type, looking for improvements in both the performance and aesthetic aspects of the form.

The structural success of the form can be understood by comparing it against the moment diagram for a three-hinged arch (Figure 1-17). The three-hinged arch creates maximum moments at its quarter points, exactly where Maillart had placed most of his material. In other words, the bridge is thick where it has to be, and thin everywhere else. It demonstrates how it works structurally. People respond aesthetically to

FIGURE 1-16 *Tavenasa Bridge over the Vordor Rhine River, Tavenasa, Switzerland, 1905, the prototype for Maillart's three-hinged arches.*

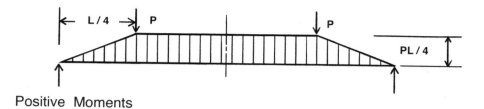

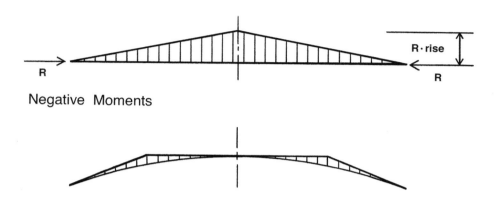

Positive Moments

Negative Moments

Combined Moments (drawn on arch axis)

FIGURE 1-17 *Bending moment of three-hinged arch with concentrated loads at the quarter points.*

that demonstration, which is why Maillart's bridges are judged a success on aesthetic grounds as well as on performance and cost grounds.

The most familiar of Maillart's three-hinged arches is the Salginatobel bridge (Figure 1-18). In this bridge he further developed his ideas, clearly differentiating the side spans. The bridge was selected by New York's Museum of Modern Art as an outstanding example of structural art in 1949 and has appeared in many books on modern art and architecture. In 1993, it was named an international engineering landmark by the American Society of Civil Engineers.

FIGURE 1-18 *Salginatobel Bridge, near Shiers, Switzerland, 1930, Robert Maillart's masterwork of structural art.*

FIGURE 1-19 *Aare River Bridge at Vessy, Switzerland, 1935, Maillart's end point in developing the three-hinged arch.*

But even with the Salginatobel Bridge Maillart was not satisfied. He decided, on aesthetic grounds, that the curve of the arch soffit was not satisfactory. He felt that there should be a noticeable break in the line of the underside of the arch to mark the presence of the center hinge. So, on his last bridges, he made the undersides of the arch segments almost straight, as shown in the bridge over the Aare at Vessy (Figure 1-19). He also continued to refine the form of the side spans. His constant goal was for improved structural performance, appearance, and economy, all at the same time. As Billington puts it, "Maillart began with aesthetics, and then, with that basic consideration in mind, looked for the best structure, where 'best' meant minimum materials, minimum cost, and minimum applied decoration."[6]

Of course, engineers always try to improve performance with no increase in cost, and often succeed. Maillart was different in that he also always tried to improve appearance with no increase in cost, and he often succeeded.

Objections are sometimes raised to a proposed aesthetic feature on the grounds that it may compromise structural performance. The question should be, does the feature really compromise the performance of the structure as a whole? Engineers do not ask for an infinite factor of safety on any structure. It would cost too much and nothing would be built. For any given structure and for any given technology, engineers arrive at an acceptable factor of safety based on the predictability of the loads and the consequences of failure. Once the safety factor is decided upon, the engineer tries to apply it uniformly across the structure as a matter of economic efficiency. There is no point in spending additional money on a feature if it will not make the structure, as a whole, safer. Similarly, once the appropriate factor of safety is decided upon, the best possible appearance should be sought within this criteria. Performance should not be used as a reason to rule out a feature that would improve appearance, even though the feature would create no reduction in the performance of the bridge as a whole. Mediocre appearance imposes its own dysfunctions, in community dissatisfaction and general ugliness. Why incur those burdens to achieve something that is not necessary?

AESTHETIC QUALITY AND COST

Many engineers point to cost or cost-related issues, such as lack of design time or lack of political support, as excuses for not paying attention to appearance. Their attitude stems from a common belief that an improvement in appearance will automatically cost more. This is not true.

Potential Attractiveness of Structural Efficiency

As explained previously in "Five Fundamental Ideas," good appearance is often associated with bridges that efficiently respond to the flow of forces in the structure. They are thick where the stresses are highest, and thin where the stresses are lowest. Because such efficiency translates into a minimum amount of material, these structures are usually economical in cost. Thus, it is often true that a good-looking bridge is also a low-cost solution.

Indeed, some of the most attractive bridges of modern times were built because they were the most economical solution to the problem. The Swiss have a tradition of selecting bridge designs through competitions where cost and appearance are both criteria. All of the bridges Robert Maillart built as he developed his ideas for three-hinged arch bridges as just described were the least expensive solutions proposed for their sites.

The design for Virginia's Maury River Bridge (Figure 5-44, page 170) reduces the spans of the main girders by almost half. The result is not only a significant reduction in cost, but a striking appearance, as well. The bridge strides across its wooded valley with the grace of a steel greyhound.

Thus, it is not necessary to spend more money to achieve a good-looking bridge. Indeed a tight budget can and should act as a spur to creativity, encouraging a search for new approaches with both cost and aesthetic advantages. Improvement in appearance should be considered in the same way as an improvement in any other area of concern: safety, durability, or maintainability. In these other areas improvements can be made that do not add cost, and may even save money. The same is true of improvements in appearance. In all of these areas the constant challenge is to find ways of making improvements without spending more money.

If the improvement does cost more money, then the question becomes: Is the improvement worth the increased cost?—keeping in mind that the bridge may be a feature of the landscape for a century or more. That question is discussed in "Costs of Responding to Larger Public Objectives" next.

While it is not necessary to spend more money to achieve a good-looking bridge, the principle, unfortunately, does not work in reverse. Not every low-cost structure will be beautiful. One has only to travel the average American freeway to confirm that economy does not guarantee beauty.

Engineering problems present many solutions. No matter how objective an engineer may be, he or she must still make some decisions where there is a degree of

FIGURE 1-20 *At Waterloo Bridge, London, long spans were required to provide space for a civic promenade, book market, and other uses.*

FIGURE 1-20 *At Waterloo Bridge, London, long spans were required to provide space for a civic promenade, book market, and other uses.*

uncertainty, and must still make choices when alternative designs perform equally well or cost the same. It is in these decisions that the differences between beautiful and ordinary will be found.

Costs of Responding to Larger Public Objectives

Bridges are prominent features of many landscapes. They are justifiably called on to meet objectives beyond their transportation function. Engineers must ask themselves whether their cost analysis is really addressing all of the dimensions of the problem, including objectives that may not express themselves in the form of number of lanes, minimum clearances, and other physical criteria.

One frequent requirement is to recognize larger patterns of activity in the area around the bridge. A bridge near a civic center, for example, may benefit from longer spans than would normally result from transportation needs alone. The longer spans may open up views to an important civic monument or scene, or permit the easier passage of crowds that gather to attend parades and civic celebrations. The Clearwater Memorial Causeway described in Chapter Six is an example in which civic views were important. Bridges must frequently respond to the public's aspirations to achieve a specific visual quality for an area or a desire to create a symbol for their community: a "signature" bridge.

Once articulated by the public authority in charge of the project, such requirements become legitimate public goals and, therefore, legitimate uses of public funds. Once such goals are established, only alternatives that meet the goals should be included in the cost analysis. Meeting such goals may indeed cost more. If the responsible public authority has decided that aesthetic quality does indeed have value, it is their right and responsibility to provide the funds to achieve it.

Citizens often request a specific historical look for their area. Usually, they unfavorably compare plain, contemporary bridges to a historic bridge in their area—a historic bridge with the detail and ornament of an earlier era. They feel that contemporary bridges, such as common parallel flange girder bridges, cannot carry the level of interest and the degree of symbolism of the older bridge. It is not just the first distant view and the view of the overall structure that they are responding to. The views from close up and at oblique angles also attract and interest people and give them an additional level of perception to explore. If their speed and location permit, people enjoy details and materials that can be appreciated close at hand, something that standard contemporary bridges rarely provide. This is a failing that modern engineering shares with other areas of modern design. As Rudolph Arnheim says:

The disorder and rapidity of modern living call for stimuli of split second efficiency. This makes for a rigid geometry, orderly enough, but too impoverished to occupy the human brain, the most differentiated creation of nature[7]

Modern architecture has a similar reputation. The early modern architects sought aesthetic appeal in the clear display of well-proportioned and carefully detailed structure. However, they also recognized people's interest in visual richness by providing thoughtful detail, appealing color, and interesting materials. In his icon of modern architecture, the Barcelona Pavilion (Figure 1-21), Mies van de Rohe designed an exposed structural system that had the elegance of a fine watch. But he also included walls of polished onyx and furniture of fine leather. His imitators compromised on the structural details and left out the onyx and the leather, and then wondered why people became quickly bored with their buildings. Architecture now faces its own demands for buildings that evoke the interest of historic structures.

Before proceeding with a request for a replica or emulation of a historic bridge, the engineer should provide the public with images of excellent contemporary structures with an equivalent of detail to determine what they really want. Often what they are saying is what they *don't* want. They don't want an ordinary girder bridge

FIGURE 1-21 *Exterior of Barcelona Pavilion. Modern architecture that allowed for elegance and interest.*

with multicolumn piers and minimal sidewalks. They do want something with an attractive, memorable shape that is friendly to pedestrians. Since almost all of the modern bridges they have seen fail to deliver these qualities and almost all the historic bridges they have seen do, their first thought is to ask for a historic imitation. Examples of contemporary bridges that combine modern structural methods, memorable shapes, and pedestrian-friendly details often will convince a community that their request for memorability and visual interest equivalent to historic bridges can be satisfied with a contemporary bridge. The wide circulation of images of the new Sunshine Skyway (Figure 1-3, page 2) showed how effective this approach could be. Soon every community with a bridge to build wanted its own cable-stayed bridge, regardless of the appropriateness of the site.

Many times, with creativity on the part of the designer, the public's desire for additional detail and interest can be met without a significant increase in cost, and the engineer should endeavor to do so. However, at some point, these criteria may require—and at the same time legitimize—additional cost, over and above what the transportation function alone would indicate. Some agencies codify an additional budget for such situations. The California Department of Transportation (Caltrans), for example, allows an additional 5 percent for significant bridges, and an additional 15 percent for "major" bridges.[8] The Federal Highway Administration has indicated a willingness to approve cost increases of as much as 15 percent for significant bridges.

This raises the question: What does the public get for the additional money? Lack of money is often given as an excuse to build an ugly bridge. It is not. Conversely, however, a big budget does not guarantee a beautiful bridge. It is possible to build a cable-stayed bridge in the wrong location and end up with a structure that towers over its neighbors and is quickly labeled an eyesore. Likewise, it is possible to smother an ordinary bridge with decorative materials, only to wonder 10 years later why it was ever attempted. To use an ice cream analogy, one can add an orgy of toppings to fine vanilla ice cream and produce a result that few would find appetizing. Having money doesn't mean that the discipline of economy and the criteria described in this book can be ignored. It means that they must be carefully applied so that the engineer can get the best possible improvement in aesthetic quality for the additional dollars, the best aesthetic bang for the buck.

The "Determinants of Appearance" section in Chapter Three gives some sense of priority: where the money will have the most positive effect. Often, it will be not in surface add-ons but in the design of the structural members themselves, with longer spans, higher clearances, more graceful shapes, or by introducing into an area a structural type not previously seen there.

The new bridge recently completed over the Severn River in Annapolis, Maryland, adjacent to the U.S. Naval Academy, shows the results of that approach. The design was selected as the winner of a competition in which both price and appearance were criteria, but in which some additional cost for aesthetic improvements was expected. The designer chose a haunched steel-box girder with flared columns. There are only two box girders, so that the slab overhangs and center span are substantial, requiring a 15-inch slab depth at some points. The bracing of the box girders is accomplished by steel box diaphragms at the piers only. Special lighting of both roadway and bridge is included.

FIGURE 1-22 *U.S. Naval Academy Bridge, by Thomas P. Jenkins, achieves aesthetic quality with longer spans and carefully shaped structural members.*

The result is a striking bridge with substantial shadow lines and an interesting underside. The impact lies in the shape of the structural members themselves, because that is where the designer chose to put the additional money.

Unusual detail, colors, and materials have their place, particularly in pedestrian areas, but the rule should be to use them in ways that enhance the overall structure, and in ways consistent with the nature of the material itself. For example, a pattern incised in the concrete might be used to accentuate the point at which forces are transferred from one member to another; or a projection might create a shadow line that accentuates the overall shape of a structural member. See Chapter Five for examples.

A town's desire for a signature bridge is not an invitation to extravagance. Signature bridges hold engineers to the highest standards of efficiency, economy, and elegance, and at the same time create opportunities for fresh thinking and new concepts.

Cost Estimating That Misleads

Improvements in appearance are often rejected because of misleading cost estimating. For example, sometimes people say a feature costs more because it is an aesthetic improvement when, in fact, it costs more because it is a change from the established practice. They mistake the cost of innovation for the cost of aesthetic improvement.

Any change, undertaken for any purpose, may well cost more the first time it is tried. Contractors working for a given transportation agency have accumulated habits, tools, and equipment suitable for current standard designs and details. Thus, the standards will often be bid for less, regardless of the intrinsic cost savings of a new proposal. However, should a new proposal with an intrinsic cost advantage be instituted, within a short time, the contracting industry will adapt to it, and the cost advantages will be realized.

As an example, 50 years ago, all plate girders were riveted. Asking for a welded plate girder was asking for a cost premium, whether the request was made for weight or aesthetic reasons. But when enough engineers asked for welded plate girders, the industry became familiar with the technique and tooled up for it. Its inherent advantages prevailed, and riveting passed out of use. Now, ironically, asking for a riveted plate girder means paying a cost premium.

Improvements made for aesthetic reasons must be given enough time to prove themselves. Special efforts must be made to ensure that builders fully understand the improvement, the reason for it, and how to build it. Contractors also like to see an improved product in which they can take pride. Once they understand the reason for the improvement, they will have positive suggestions for how to do it better. This enthusiasm will contribute to both economy and quality.

Then there are procedural traps that can mislead engineers who are trying to estimate the costs of improvements made for aesthetic reasons (and other reasons, too). Here are some frequent cases:

- *Suboptimization trap.* All of the pluses and minuses of a change have to be added before a true picture is realized. For example, reducing girder depth may increase the cost of the girders themselves; but the reduction in required vertical grades may save more in approach roadway and right-of-way costs.

- *Unit cost trap.* Often costs are compared on the basis of unit costs of material—so many dollars per cubic yard of concrete or pound of steel. This tends to give a misleading picture when there are significant differences in the alternatives beyond the quantities of material involved. For example, a decision to use a two-column pier in place of a three-column pier might involve more concrete in the pier. But, it might also produce offsetting savings in forms, footings, and placement of reinforcement because of the reduced number of columns.

- *Precision trap.* Construction estimating is an imperfect art. It is a rare day when the engineer's estimate comes within 5 percent of the contractor's bid. The prices contractors bid for items change from day to day, depending on the size of their workload, the costs of materials, the availability of labor or equipment, and their degree of knowledge. Any feature or combination of features that costs less than 5 percent of the total cost of the bridge is essentially outside the range of precision of cost estimating. It might as well be treated as cost-neutral, given the inability of the designer to predict its eventual effect on the total price of the bridge.

Wrong Standard Details

Standard designs and standard details can be both aesthetic problems and aesthetic opportunities. Many large bridge-building agencies build so many structures that, for efficiency's sake, they have developed libraries of standard designs and details. These are often pointed to as a reason not to undertake an aesthetic improvement, on the grounds that the nonstandard improvement would automatically add cost. However, the benefits of standardization are as available for attractive bridges as for ordinary bridges.

Caltrans has built generations of attractive bridges based on standard elements. In the opinion of James E. Roberts, the agency's chief bridge engineer, these are the least expensive bridges they can build.[9] All of the benefits of standardization apply. California contractors are familiar with the structures; they have accumulated stocks of standard forms and fittings. And the product is outstanding structures. (See Chapter Seven for more on this subject.)

It's easy to understand the need for standardization to meet production needs. The question is, why not standardize beauty instead of mediocrity? Why not grind out quality?

Design Time or Cost

The perceived need for additional design time is often given as a cost-related excuse to build ordinary bridges. Naturally, the first time designers try to apply aesthetic ideas or techniques, it will take additional time. As in any other situation, there is a cost for innovation. There is a learning curve to be faced; new computer-aided design (CAD) techniques, training, and some experimentation must be expected.

However, once a design staff is skilled in issues of appearance, is equipped with the necessary design tools (such as 3D CAD) and has access to attractive standard details, it can turn out a good-looking bridge as quickly as an ugly one. The staff may, in fact, save time by avoiding the delay that occurs when objections are raised to an ordinary design. Why not do it right the first time?

THE BRIDGE ENGINEER'S ROLE

Because the appearance of bridges is dominated by the shapes and sizes of the structural members, appearance cannot be an afterthought or an add-on to bridge design. Concern for appearance must be an integral part of engineering. It is a difficult challenge. As Billington interprets Maillart's view of it:

> The bridge art is, therefore, vision disciplined by technique; and more specifically, a vision of the public landscape formed by economic constraints on public structures. It is a difficult art, with the artist continually struggling to control his elements in the face of public opinion, codes, budgets, and politics.[10]

Many engineers see themselves as a type of applied scientist, analyzing structural forms established by others. In the case of most everyday bridges, the selection of form is based largely on precedents and standards established by the bridge-building agency. For example, the client agency may specify that the form of a highway overpass has to be a welded plate girder because that is the agency's preference, or because there are particularly cost-effective local steel fabricators, or even because the steel industry is a dominant political force in that state. Whether a welded plate girder is in fact the best form for that particular site may not even be considered.

Seeing oneself as an applied scientist is an unfortunate state of mind for a design engineer. It eliminates the imaginative basis of design and forfeits the opportunity for the integration of form and structural requirements that can result in structural art. Design must start with the investigation, selection and proportioning of the structural form. All that follows, including the aesthetic impression the bridge makes, will depend on the quality of the form selected. All of the potential for creative structural art begins with that decision. This decision can be made well only by the engineer because it must be based on a knowledge of the forces involved and the forms best suited to handle them.

Creative engineering design consists not in applying free visual imagination alone, nor in applying rigorous scientific analysis alone, but in applying both simultaneously. In the words of the Spanish engineer Eduardo Torroja:

> *The imagination alone can not reach such (elegant) designs unaided by reason, nor can a process of deduction, advancing by successive cycles of refinement, be so logical and determinate as to lead inevitably to them. . . .*[11]

The art starts with a vision of what might be. The development of that vision is the key, for it sets the goals for all that follows. The vision assumes reality through a process that many engineers call "conceptual engineering." It is the stage at which all plausible possibilities, and some not so plausible, are examined in sketch form. The examination must include, at a rough level of precision, the whole range of considerations: performance, cost, and appearance. This important stage is often ignored or foreclosed, based on preconceived ideas or prior experience that may or may not apply.

The reasons often given for short-changing this stage include: "Everybody knows that _____ is the most economical structure for this location," or "We always build _____ in this state," or "Let's use the same design as we did for _____ last year." When these thoughts are the starting point, it is unlikely that the most promising ideas will ever emerge. Instead, there is a premature assumption of the bridge form, and the engineer will move immediately into the analysis of the assumed type.

That is why so many engineers mistake analysis for design, when design is more correctly the selection of the form in the first place, and is by far the more important of the two activities. Before there is any analysis, there must be a form to analyze.

Engineers also focus on analysis in the belief that the form (shape and dimensions) will be determined by the forces as calculated in the analysis. But, in fact, there are a large number of forms that can be shown by the analysis to work equally well. It is the engineer's option to choose among them, and in so doing he or she will determine the forces by means of the form, not the other way around.

Take the simple example of a two-span girder bridge using an existing bridge, MD 18 over U.S. 50 (Figure 1-23.1). Here the engineer has a wide range of possibilities. He or she can give the girder parallel flanges, or give it a haunch of a wide range of proportions (Figure 1-23.2). The moments will depend on the stiffness at each point, which in turn will depend on the presence or absence of a haunch and its shape. The engineer's choice of shape and dimensions will determine the moments at each point along the girder (Figure 1-24). The forces will follow the choice of form. Within limits, the engineer can direct the forces as he or she chooses.

FIGURE 1-23.1 *Maryland 18 over U.S. 50, one approach to a two-span bridge. (See color insert.)*

FIGURE 1-23.2 *Comparative computer-drawn variation of Maryland Route 18 over U.S. 50.*

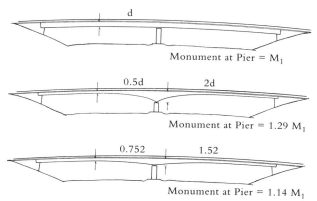

d

Monument at Pier = M_1

0.5d   2d

Monument at Pier = 1.29 M_1

0.752   1.52

Monument at Pier = 1.14 M_1

FIGURE 1-24 *Moments determined by the choice of form, (calculated based on a typical noncomposite two-lane highway overpass with four girders and two equal 110-foot spans).*

FIGURE 1-25 *Arenal Bridge, Cordoba. A form invented with the goal of structural efficiency. The concrete shape acts compositely with the steel box to handle the high compressive forces in the negative moment area.*

Now let us examine which of the forms in Figure 1-24 the engineer should choose. All of them can support the required load. Depending on the specifics of the local contracting industry, all of them may be essentially equal in cost. All of them would perform equally well, and all of them are comparable in cost, leaving the engineer a decision that can only be made on aesthetic grounds. Why not pick the one that he or she feels looks best?

That, in a nutshell, is the process that all of the great engineers have followed. Quoting Billington, again in regard to Maillart:

The engineer cannot choose form as freely as a sculptor, but he is not restricted to the discovery of pre-existing forms as the scientist is. The engineer invents form, and Maillart's career shows that such invention has both a visual and a rational basis. When either is denied, then engineering design ceases. For Maillart, the dimensions were not to be determined by the calculations, and even the calculation results could be changed (by adjusting the form) because a designer rather than an analyst is at work. Analysis and calculation are the servants of design.[12] [See Figure 1-25.]

At this point some will protest that other considerations (such as costs or the preferences of the local contracting industry) will indeed differentiate and determine the form. Too often that belief is based on unexamined assumptions, such as, "The local contracting industry will not adjust to a different form," or "Cost differentials from (a past project) still apply," or "The client will never consider a different idea." Or that belief is based on a misleading analysis of costs, as described in the previous section. Or that belief may be simply habit—either the engineer's or his or her client's—often expressed in the phrase "We've always done it that way." At its base, the role of the engineer is to question those assumptions and beliefs, including his or her own, for each structure attempted. Simply accepting these assumptions and beliefs places an unfortunate and unnecessary limitation on the quality of the resulting bridge for, by definition, improvements must come from the realm of ideas not tried before. As Captain James B. Eads put it in the preliminary report on his great bridge over the Mississippi River at St. Louis, *"Must we admit that because a thing has never been done, it never can be, when our knowledge and judgment assure us that it is entirely practicable?"[13]*

FIGURE 1-26 *Shear connectors assure composite action between the steel and concrete portions of the webs of the Arenal Bridge.*

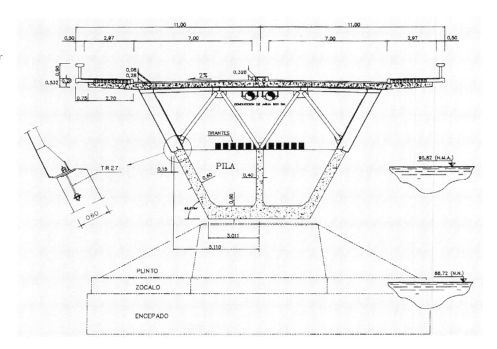

The art of questioning those assumptions and beliefs will produce the open mind that is necessary to develop a vision of what each structure can be at its best. The engineering challenge is not just to find the least costly solution, but to bring forth elegance from utility. We should not be content with bridges that only move vehicles and people. They should move our spirits as well.

USING THIS BOOK

This book aims to help engineers improve the engineering of their bridges. Appearance will be a major concern, but not the only concern. Judgments about appearance

FIGURE 1-27 *Engineering elegance in Seville, Spain, the Puente de la Barqueta.*

must always be made in the context of judgments about performance and cost, and this book concentrates on those relationships.

Chapter Two develops basic principles of aesthetics and perception in the transportation environment. The chapter ends with a list of 10 features of all bridges (superstructure, piers, parapets, etc.) according to their importance in creating the bridge's aesthetic impact. These elements are labeled the "Determinants of Appearance." Chapter Three suggests procedures to apply these basic ideas to bridge design.

Chapters Four and Five look at each Determinant of Appearance. The chapters present specific guidelines for each determinant. The guidelines are presented not as absolute rules but as indications of what has worked well in previous structures. They are presented in the form of comparative sketches so that the engineer can decide for himself or herself how well they apply to the bridge in question. The sketches are characterized as either "ordinary," representing problem designs, or "better," representing markedly improved designs. The examples are accompanied by analyses of the underlying visual principles.

The guidelines incorporate a consensus opinion about how good appearance can be achieved. The consensus has been derived from analysis of the visually successful bridges of the modern era. It emphasizes simplicity, apparent slenderness, horizontal continuity, and the inherent attractiveness of efficient structural forms. The consensus has developed over several generations of experience with modern structures.

Chapter Six shows examples that incorporate the guidelines to a greater or lesser degree, using actual bridges from typical locations. Again, bridges are characterized as "ordinary" or "better." However, "better" is not "best." The better bridges are shown as targets to surpass, not models to copy. Every bridge is unique, and only its designer can recognize which of these ideas may apply, which must be adjusted, and when new ideas must be developed. "Best" is that unique design that precisely fits the situation at hand, using the best creative thinking of the designer to simultaneously solve the problems of performance, cost, and appearance.

Engineers should use these chapters as thought provokers, not thought inhibitors. They should be used as tools in observing bridges, and in making personal judgments about what works well and what does not.

Chapter Seven looks at the aesthetic responsibilities of bridge-building organizations, the role of citizens and elected officials, the responsibility of professional engineering organizations and the role of engineering educators in improving the aesthetic quality of bridges. Finally, this chapter examines the great potential of engineering design competitions not only to select excellent concepts for important bridges, but also to improve the design of bridges everywhere.

---

1   Schuyler, Montgomery, May 1883. "The Bridge as a Monument." *Harper's Weekly* 27: 326, reprinted in *American Architecture and Other Writings,* ed. William H. Jordy and Ralph Coe, New York: Athenum, 1964, p. 164.

2   Survey by the Federal Highway Administration on Context Sensitive Design.

3 Petroski, Henry, September–October 2002. "Benjamin Franklin Bridge." *American Scientist* 90; see also Petroski, Henry, 1995. *Engineers of Dreams: Great Bridge Builders and the Spanning of America.* New York: Alfred A. Knopf.

4 Billington, David P., 1983. *The Tower and The Bridge, the New Art of Structural Engineering.* Princeton, NJ: Princeton University Press.

5 This is a summary of Maillart's thought process as described in Billington, David P., 1979. *Robert Maillart's Bridges: The Art of Engineering.* Princeton, NJ: Princeton University Press.

6 *Robert Maillart's Bridges,* p. 117.

7 Arnheim, Rudolph, 1969. *Visual Thinking,* Berkeley: University of California Press.

8 Gloyd, Stewart, February 1994. "California: A Quantified Bridge Aesthetics Case Study." *Concrete International.* Farmington Hills, MI: American Concrete Institute.

9 Roberts, J.E., 1990. *Esthetics in Concrete Bridge Design,* ed. S.C. Watson and M.K. Hurd. Detroit: American Concrete Institute.

10 Billington, David P., 1989. *Robert Maillart and the Art of Reinforced Concrete.* Cambridge, MA: The MIT Press.

11 As quoted in Billington, David P., 1982. *Thin Shell Concrete Structures.* New York: McGraw-Hill Publishing Co.

12 Billington, David P., 1989. *Robert Maillart and the Art of Reinforced Concrete.* Cambridge, MA: The MIT Press.

13 As quoted by Petroski in *Engineers of Dreams.*

chapter two

UNDERSTANDING THE BASICS

"There is a kind of human authority which needs no uniforms or symbols to make itself evident, but which makes a man stand out in a crowd simply because his movements and posture portray a powerful, controlled, and self-assured personality."[1]

—SINCLAIR GAULDIE, BRITISH ARCHITECT

TERMINOLOGY

Aesthetic reactions are created by the eye/brain at the moment an entity is seen. They are not created by words. Thus words cannot completely reflect the phenomena that we are trying to describe and evaluate. Nevertheless, words are a necessary part of our communication and are often used to describe or explain an aesthetic reaction after it has occurred. In order to use words to communicate about aesthetics we need a commonly understood terminology.[2] The following terminology is borrowed from other visual design fields and applied to bridge and highway design.

Visual Aspects

To be able to talk about an object, it is helpful to have names for its visual characteristics. "Aspects" are the visual characteristics of the visible parts or elements of a bridge or bridge/site composition.

Line[3]

A line is a direct link between two points, either real or implied. The strongest lines on a highway are created by the pavement edges. Other prominent lines are created by railings, girders, piers, abutments, and the top edges of retaining walls and noise walls. (See Figures 2-1.1 and 2-1.2.)

Shape

When a line encloses an area, it creates a two-dimensional surface with spatial dimensions of height and width. Shape is the configuration of a two-dimensional surface.

FIGURE 2-1.1 *Our aesthetic reaction to this bridge is strongly influenced by the attractiveness of its parallel curvilinear lines.*

FIGURE 2-1.2 *The jagged line of the top of the noise wall contrasts unattractively with the smoothly curved line of its base and the even more gentle curves of the highway alignment.*

Form

Form is the configuration of three-dimensional objects, adding depth to the height and width of shape. The visual experience of moving under or over a bridge is primarily influenced by the form of the bridge, its horizontal alignment, vertical profile, span arrangement, width, depth, and relationship to adjacent structures. The form of a bridge is seen contained within the space that creates its environment. (See Figure 2-3.)

Color

Color, for our purposes, is the ability of a surface to reflect light of a specific spectral composition. Color can be applied to define, clarify, modify, accentuate, or subdue the visual effects of line, shape, or form. Colors are perceived differently at different times of the day and at different times of the year because of changes in light conditions created by the position of the sun and atmospheric conditions. Colors are

FIGURE 2-2 *The haunch provides a more interesting and attractive shape than a girder with parallel edges. It also demonstrates how the girder works, making the girder thicker where the forces are greatest.*

FIGURE 2-3 *The three-dimensional form of a bridge is the result of the interaction of the parapet, girders, piers, abutments, and surrounding topography.*

also influenced by the background against which they are seen, and their appropriateness is often judged in terms of their fit with their background. Background is particularly important for most highway color selections because the highway element is almost always a small part of a much larger scene, the colors of which are outside the designer's control (Figure 2-4).

Texture

Texture is a characteristic of the surfaces of objects and describes their roughness, which may be patterned or random. Texture helps define form because the roughness causes mini-shadows on the different surfaces of an object, depending on their angles relative to the light source, and thus helps make some parts of an object appear

FIGURE 2-4 *Bright color clearly differentiates the elegant load-carrying tubular arch from the deck it supports. Jörg Schlaich's Pragsattel I pedestrian bridge, Stuttgart, Germany. (See also color insert.)*

FIGURE 2-5 *Even precast retaining wall elements read as texture at the distances and speeds typical of the highway environment. Unfortunately, the lines created by the hexagonal pattern conflict with the sloped lines of the parapet and the vertical lines of the abutment.*

FIGURE 2-6 *Orienting surfaces to create areas of shade and shadow.*

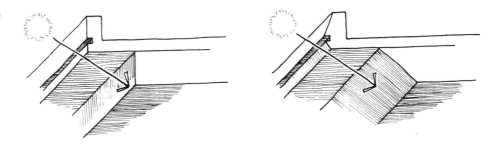

darker than others. Texture can be used to soften or reduce imposing scale, add visual interest, and to introduce human scale to large objects such as piers, abutments, and retaining walls. Distance and motion alters the perception of texture. When viewed from a distance or at high speeds, fine textures blend into a single tone and appear flat. As a general rule, the greater the distance, the higher the observer's speed or the larger the object to which it is applied, the coarser or larger the texture must be (Figure 2-5).

Shade and Shadow

Areas of comparative darkness caused by the interception of light by intervening parts of the bridge or another nearby object. Shadow is the term usually used when the intervening object can easily be identified. The shadow that the deck overhang of a girder bridge casts on the outside girder can be a very strong component of the appearance of the bridge (Figure 2-6).

Reflections

Images of a bridge visible in the water below can be a very important part of the impression made by a bridge (Figure 2-7). Reflections of light from the ground below or other nearby objects can also help illuminate the underside of a bridge and influence our impression of it.

FIGURE 2-7 *Reflections mirror the bridge more or less clearly, depending on the stillness of the water. They can make the bridge look taller. Brainerd Bypass over the Mississippi River, Brainerd, Minnesota.*

FIGURE 2-8.1 *This bridge has a little bit of everything and thus has no order.*

FIGURE 2-8.2 *The repetitive pier shapes and continuous girder give this bridge a sense of order.*

Visual Qualities

Visual qualities describe the arrangement of the visible aspects of an object and are used to evaluate a visual composition. Visual qualities are intangible; they are perceived characteristics that exist only in the mind of the evaluator.

Order

Order is the arrangement of design elements so that each element has a proper place and function and the whole works together as a unit without confusion (Figures 2-8.1 and 2-8.2).

Proportion

Proportion is a method of creating a sense of order by assigning appropriate relative sizes to the various elements. The goal is to create visually pleasing ratios between the various parts of a structure: between the height, width, and depth of its parts; between solids and voids; between areas of color and texture; between areas of sunlight and shadow; and between parts of the structure that are thin and parts that are thick (Figure 2-9). Proportion can suggest the order of significance of the elements or the role played by the elements in a structure—their relative size classifies some as performing principal functions and others as attending to secondary functions. For example, a slender column suggests a light load-carrying function, whereas a thick column suggests the opposite.

Rhythm

Rhythm is a method of creating a sense of order by repeating similar elements in, on, or around a structure. The goal is a natural flow that is satisfying to the eye. It requires that the elements have some similarity of visual characteristics in addition to a modulated placement. In bridges, major rhythms are created by the repetition of similar pier shapes. Minor rhythms may be created by the spacing of light poles, post spacing within a railing, or even the horizontal rustication on a pier (Figure 2-10).

FIGURE 2-9 *This rigid frame bridge creates an impression of slender grace because of its proportions, the comparison of the large depth at the abutment to the shallow midspan.*

Harmony

Harmony means that aspects and elements of a design have visual similarity. If forms, shapes, or lines in a design have more dissimilar characteristics than they have similar characteristics, they are not likely to be perceived as harmonious (Figure 2-11).

Balance

Visual balance is the perceived equilibrium of an object around an axis or focal point. Rather than a physical balance, it may refer to an equilibrium of visual aspects of a design, such as size, apparent weight, color, or texture (Figures 2-12.1 and 2-12.2).

Contrast

Contrast creates visual interest by complementing the characteristics of some aspects or elements with their opposites. This adds a heightened awareness of each other. Contrast often takes the form of dramatic differences in color or in light and

FIGURE 2-10 *The large main span creates a variation in the otherwise consistent major rhythm of the piers, while the light posts provide a regular minor rhythm. U.S Naval Academy Bridge, Annapolis, MD.*

FIGURE 2-11 *The shapes of the piers are similar, and thus harmonious, even though their sizes are different. I-95 and US 50, Cheverly, Maryland.*

shadow. Contrast is often used to create dominance, where one of two contrasting elements commands visual attention over the other. One becomes the feature and the other becomes the supporting background. A dominant theme is essential in organizing the design into a pleasing aesthetic experience (Figure 2-13).

Scale

Scale refers to the size relationship between a bridge, its various features, and the highway and its surroundings. We judge most things by their usefulness to us, which requires a comparison of the thing to the size of the human body. We are all familiar with windows and doors and thus can easily judge the size of most buildings by counting the windows and doors. When we feel that an object has an appropriate size relationship to the size of the human body, we say it has a human scale.

FIGURE 2-12.1 *This bridge is visually balanced about the highway centerline. Each end is balanced over its pier as well, physically as well as visually.*

FIGURE 2-12.2 *The end is tied down at the abutment to allow minimal depth at the midspan. The overall impression is a bridge of incredible thinness, actively balanced over its supports.*

FIGURE 2-13 *This pedestrian bridge by Jörg Schlaich creates its visual impact in part by the contrast between the thick concrete girder and the tenuous steel cables holding it up. Main/Danube Canal, Kelheim, Germany.*

FIGURE 2-14 *Bridge scale versus human scale. The two pedestrians are dwarfed by the bridge overhead and its piers. The pedestrian ramp helps mediate the difference by introducing an element of human scale into the scene.*

Because they must span long distances and are built for vehicles moving at high rates of speed, bridges and highways are much larger in size than all but the largest buildings. Bridge elements such as piers or girders can be very large, but appear "in scale" with the highway environment. Conflicts in scale become apparent when highway elements become part of a pedestrian environment or adjoin buildings (Figure 2-14). Then ways must be found to reduce the apparent size of the highway element so that it fits into the smaller-scale environment.

Unity

Unity provides the observer with a sense of wholeness. It is generated by some central or dominating perception in the composition. It encompasses all the other qualities, and it refers to the combined effects of all other aesthetic qualities applied

FIGURE 2-15 *One way to achieve unity is to design all elements with the same family of shapes, with the same standard details and the same colors. The Big I Interchange, Albuquerque, New Mexico. (See also color insert.)*

simultaneously. Unity is the condition in which all of the elements, aspects, and qualities of the bridge are in accord, thus producing an undivided total effect (Figure 2-15).

ANALYZING THE APPEARANCE OF BRIDGES

We perceive bridges primarily through the eyes. That means that success in bridge aesthetics depends first on recognizing the effects of light, shade and shadow, color, and visual illusion. It also means that it is necessary to determine the positions and characteristics of likely viewers of the bridge. Viewers in cars are a special case because they have a more predictable and limited range of views, which tends to focus their attention on bridges. Our appreciation of a bridge is influenced to some degree by what is in view around it. Thus, the openness, motion, and horizontality of the typical highway environment will affect how bridges appear.

Finally, accurate analysis of the appearance of bridges can be done only in the field or through media that represent the three-dimensional reality of the structure: perspective drawings, models, and photographs.

Perception of Bridges

Let us begin with the fact that we perceive a bridge primarily through our eyes. Usually, we are too far away for any other sense to come into play. Often we are in a vehicle and perceive the bridge through the windows of that vehicle. Exceptions exist where persons on foot can approach some part of the bridge, such as the columns of a viaduct over a city street. In these cases, the sense of touch, the feeling of safety, or the noise and vibration of vehicles passing overhead might become involved. These exceptions will be discussed elsewhere in the guidelines. The following discussion concerns visual perceptions.

Light, the medium through which visual perception takes place, is the heart of the matter. Daylight is the universal light source for bridges, and in most cases, the only relevant one. The quality of daylight changes with the time of day and time of year, the degree of haze or cloud cover, the structure's geographic location and orientation, the orientation of the viewer, the season of the year, and the overall background colors of the environment.

Early morning and late afternoon daylight produces long shadows and rich-but-less-bright colors. Mid-morning and mid-afternoon light produce average shadows and true, bright colors. Noontime colors tend to be washed out by the bright light, and shadows are at a minimum. During winter, the light fades, shadows lengthen, and colors are generally not as bright. At any time of year, haze will dull colors and shadows, so there is less difference between the brightness of surfaces in or out of shadows. Finally, cloud cover eliminates shadows and dulls colors, giving them a gray undertone.

At any given location, most combinations of clear sunlight, evening and morning light, and haze and fog will occur during the course of a year. But it is also true that certain combinations are more likely in some locations than in others, and some are unique to their locations. The brilliant light of the Florida summer sun with its deep blue sky has no equivalent in Minnesota. Conversely, the thin, clear light of a Minnesota winter with its pale blue sky has no equivalent in Florida.

Characteristics such as these are dominant effects in their locales. They represent the type of light in which bridges will most often be observed. Engineers need to be aware of the characteristic light and sky color of the area in which they are working.

The following are some observations about light in different parts of the United States. They should be confirmed by personal observations on the site.

In the Northeast and Midwest, summer brings clear sunshine and blue skies with sharp shadows; winter brings many cloudy days with indistinct shadows. Mid-Atlantic and Southeastern summer days tend to be hazy. Though there is plenty of light, the sky tends to be colorless, and the shadows are fuzzy. Winter brings clearer skies, but the sunlight is weaker. Florida has the strong light and blue sky just described. The Southwest has almost constant strong sunshine with deep blue skies. The West has similar light, though weaker in winter, depending on latitude. The Northwest's light is affected by the frequent cloud cover; it is diffuse and relatively colorless because of the gray skies. Compare the dark shadows of Figures 1.8, page 4, caused by clear Rocky Mountain sunshine where no details can be in the shadowed area, with Figure 4-17, page 109, where the diffuse British Columbia light leaves much detail visible in the shadowed areas.

Because daylight is the medium of visual perception, the orientation of a bridge is a major influence on how it will be perceived. Surfaces that face south will be consistently the brightest; their shadows will change relatively little during the day, but change significantly from season to season; their colors will tend to fade most quickly. Surfaces facing east or west will be in shade half the day and be marked by strong, rapidly changing shadows the other half. Surfaces facing north will be in shade at all times; their colors will stay bright the longest.

The background against which a bridge is seen also influences the impact the bridge will make. East Coast and Midwestern background colors tend to be multiple shades of green in spring and summer; yellows, oranges, and browns in the autumn; and brown and grays in winter. California backgrounds are bright green in the early spring and brown the rest of the year.

Designers can't control daylight and background, and can rarely control orientation, but shadow and shade are susceptible to control. Overhangs, projections, grooves, and recesses create areas and patterns of shade and shadow that help create the aesthetic impression (Figure 2-6, page 30). The brightness of surfaces can be changed by changing their orientation to the sky. Surfaces slanted toward the sky will be brighter than vertical surfaces or surfaces slanted toward the ground. The designer can also change the reflectance of surfaces, for example, by using white concrete, to accentuate the effect of light and shadow.

Nighttime illumination from headlights, or as incidental illumination from roadway lighting luminaries, is rarely sufficient to do more than pick out major shapes. Given these limitations, plus the fact that most people's nighttime highway experience is largely occupied with the difficulties of the driving task, it is probably not worthwhile to be overly concerned with the appearance of most everyday bridges at night. The major exceptions would be those circumstances where the bridge itself deserves lighting because of its place in the environment or its symbolic importance, or where the use of the bridge requires lighting the space on or below it. Such situations are discussed in Chapter Five.

Viewpoints: Fixed and Moving

When an individual forms an aesthetic impression of a bridge, where the bridge is viewed *from* will strongly affect his or her impression. If the engineer wants to control that impression, it is important to know the likely positions of viewers of the bridge. For a bridge over a park valley, for example, these would be park users on the trails below (Figure 2-33, page 47); for a viaduct over a city street, these would be pedestrians and drivers on the street. For most bridges, there are many such viewpoints. For prominent bridges, the viewing area may cover several square miles and incorporate whole communities within sight of the bridge.

The bridge itself creates new viewpoints overlooking the environment. For a bridge over a major barrier or at an entrance to a town, the act of crossing the bridge may have great symbolic importance. In the past, this was recognized by the placement of statuary, plaques, elaborate lighting fixtures, or viewing platforms on the bridge. It is not appropriate to just copy historic elements to place on a modern bridge; rather, contemporary versions that accomplish the same symbolic and functional purpose are in order.

In all cases, the view by the user of the bridge itself must be recognized. Curving approaches often create dramatic views of the oncoming structure, and there are always overhead features and the insides of parapets to be considered.

Not all viewpoints can be accommodated to the same degree; it is often necessary to assign priorities among them, perhaps giving more weight to the most numerous observers, or more weight to the view from, say, the town square than to the view from the town industrial park.

Once the viewpoints have been established, the engineer should evaluate the observers at the various positions. For stationary or pedestrian viewers, the most important variables are distance and relative elevation. For bridges viewed primarily from a half-mile away, the major concern must be for overall shapes and the colors of large areas. Details will not matter. For bridges to be viewed close at hand, and particularly bridges to be used by pedestrians and bicyclists, details and surface texture become major concerns.

Most bridges are seen primarily from highways passing beneath them (Figure 2-16); it is a very controlled situation. The viewers are moving along a prescribed line

FIGURE 2-16 *The traveler's viewpoint; also a good example of the strong shadows that can be created on a clear day by overhangs and projections.*

(their highway lane) at a constant eye height and a more or less constant rate of speed. The point at which the bridge first comes into view, the length of time that it is within view, and the size of the bridge within the visual frame at each point, can all be predicted and are the same for each observer. The visual experience is analogous to that of a movie, where the windshield is the screen, and the designer controls what is presented on each "frame."

For viewers in cars, the most important variable is speed; the second is distance to the bridge. As we travel faster and faster, two things happen to our visual perception: our field of view narrows (less and less is noticed on the periphery; Figure 2-17) and our point of focus projects further and further ahead.[7]

This is partly a result of the physics of the situation: as speed increases, the periphery of the visual field moves across the field faster than the center, until it is moving too fast for the eye/brain to process, and becomes blurred. The only thing that stays in focus is the center of attention, the highway itself. Also, we have a sub-conscious sense of stopping distance: how far ahead are the events to which we must react to *right now*. At highway speeds, those events are 300 to 500 feet or more in front of us, depending on how fast we're going, so we focus at that point.

All this means is that, at 55 mph, our last and best view of a highway overcrossing is from about 300 feet away. By the time we get to the bridge, we are looking 300 feet beyond. The bridge itself, at that point, is a blur in our peripheral vision. The parts of the bridge that we see best are those that are visible in front elevation. The undersides, the sides of the abutments, and piers are simply part of the peripheral blur. (We can, of course, make the effort to turn our head and look at an abutment wall, but even then the view will flash by so fast that few details will be recognized.)

These basics of perception also mean that, at highway speeds, the field of view in focus has narrowed to the point at which the highway itself occupies 80 percent of it, which means that the bridges are always onstage, front and center (Figure 2-18).

What are the implications of these facts? One is that any feature of the bridge

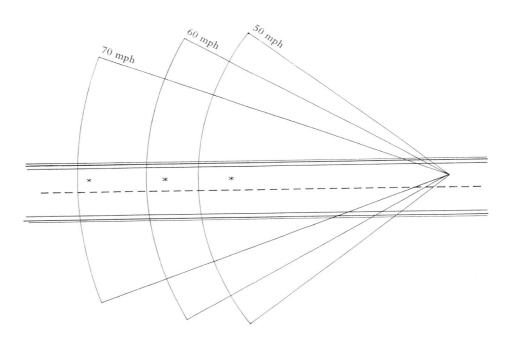

FIGURE 2-17 *The cone of vision at highway speeds.*

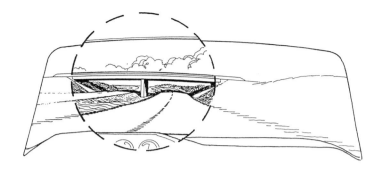

FIGURE 2-18 *The driver's area of focus at highway speeds.*

that is meant to have a visual impact must be large enough to be seen at 300 feet. A second implication is that continuous horizontal lines, parallel to the line of movement, stay in focus and are easily understood and appreciated. If pleasingly shaped, they can be major sources of enjoyment (Figure 2-19); conversely, flaws in the horizontal alignment will be jarringly evident. That's why misalignment of a median barrier is so annoying.

Vertical lines, in contrast, quickly move into the peripheral blur. A long, evenly spaced series of them will be perceived as an annoying flicker in the peripheral vision. Vertical elements that are large or close to the point of focus stand out in the peripheral blur and become prominent out of proportion to their physical position. Thus, piers and abutments close to the edge of the roadway are seen as prominent and threatening, though they may be well outside the actual physical clearance envelope. Keeping such elements well back from the roadway gives the driver visual reassurance that the space of the highway is continuous, and is not interrupted by the bridge.

Effects of Surroundings

How a bridge is seen will depend, in important ways, on what else is seen around it at the same time. People will form a different perception of a bridge if it is seen against a backdrop of skyscrapers instead of a backdrop of mountains. Many bridges are seen in the specialized environment created by highway facilities themselves with their wide rights-of-ways and large interchanges.

FIGURE 2-19 *Stone guardrails on the Baltimore–Washington Parkway establish a pleasingly curved, complementary line, which adds to the traveler's enjoyment.*

FIGURE 2-20 *Highway 404/410 Interchange, Toronto, showing large clear openings achieved with torsionally stiff post-tensioned concrete voided slabs.*

Highway Environment

The essence of the highway environment is movement. Standing for a few moments next to a freeway ramp will bring that fact home. Multiton metallic objects hurtle along curved paths at high speeds. Occasionally, groups of these objects merge and diverge in various patterns. At major interchanges, the patterns become quite complex and can involve vehicle paths crossing at multiple levels and locations. The movement may seem discontinuous, a misperception reinforced by the use of still photography to depict bridges and highways. In fact, the movement pattern remains in the mind even when vehicles are not present, exerting a kind of visual potential energy. This reality is better represented by time exposures of roadways taken at dusk showing the sweeping lines traced by moving headlights and taillights. Bridge lines and pier placements that parallel and reinforce these movement patterns will seem appropriate, and lines that are at angles to the pattern will appear jarring and out of place. A concern for safety, if nothing else, would suggest large clear openings through structures and a minimum of barriers. Where barriers are necessary, their orientation should be parallel, or at slight angles to, the lines of travel, with generous clearances.

These same features work to improve both the driver's psychological comfort and aesthetic reaction. Large openings mean that the driver can see through to the other side and know what is coming next (Figure 2-20). Large openings also mean that potentially threatening vertical lines from piers and walls are out of the field of focus, allowing lines paralleling the movement pattern to create a positive impression.

Full realization of these potentials requires early and comprehensive communication between the road designers, bridge designers, traffic engineers, and landscape architects. Opportunities to coordinate all aspects of the highway at an early stage will result in improved safety, improved appearance, and probably lower cost. Early attention to the appearance of the structures may result in slight alignment adjustments, which can often improve bridge appearance without compromising safety or cost. Early evaluation of sign design may identify safer locations for sign supports. Coordinated multidisciplinary attention can affect interchange layout as well. For example, moving a ramp gore from underneath a bridge takes it from the shadows into the light, which makes a significant difference in its visibility and safety.

Modern highway environments have been expanded to accommodate safety grading and large ramp radii. This means most highway bridges occur within a gently sloped landscaped area with the bridge itself being only a small part of the visual scene. The scene itself is predominantly spread out and horizontal. This is, of course, a matter of degree, which varies depending on the complexity of the highway and the shape of the natural surroundings. The basic point remains that the bridge itself is a relatively small object in a much larger landscape, where the dominant dimensions, compared to the bridge, are horizontal. This fact, combined with the horizontal nature of the vehicular movement that the structure carries, indicates that, in most cases, the horizontal elements of bridges should be emphasized. (See Figure 2-21.)

There are exceptions. Within a major multilevel interchange, a single, multilevel structure or a series of closely spaced and overlapping structures will be dominant enough to establish their own environment. The visual impact of this assembly needs to be studied to determine the structural/aesthetic approach that is appropriate. Figure 2-53 (page 57) shows an example in which all of the bridges are the same basic type of structure: steel box girders. This structural type accommodates varying span and clearance requirements within a single structural form. The result is visual unity, as well as a degree of openness. It is similar to the result obtained in Ontario with concrete, voided slab structures (Figure 2-20 page 39).

Natural Environment

The range of possible natural environments is immense. First, there is the question of form, ranging from the absolute flatness of a tidal inlet between coastal barrier islands to the confining cliffs of a Rocky Mountain canyon. Then there is color, which changes from place to place and from season to season. Finally, there are the differences in texture created by the presence or absence of vegetation, the varying types of vegetation, and the differences in rock and soil.

It is not possible, nor desirable, to reflect all of these differences in specific features of the bridge. However, we must be aware of the effect these features will have on the public perception of the bridge. In some cases, the selection of structural type and shape can be done in ways that reinforce the characteristics of the environment. An arch between granite canyon walls is an obvious example, which takes advantage of the natural efficiency of the arch in that situation. The Salginotobel bridge (Figure 1-18, page 13) is a classic example.

Urban Environment

Urban environments are usually more confined. Urban structures often require retaining walls and are sometimes overshadowed by buildings. Here, every visual surface is man-made and often hard-edged, and the vertical dimensions are of the same order of magnitude as the horizontal dimensions. More emphasis on the vertical may be in order. However, the continuity of the driver's line of vision is still paramount—horizontal lines should follow the highway geometry as much as possible, with as much "visual space" as possible evident to the driver. (See Figure 2-22.)

The viewpoints of pedestrians and slow-speed drivers become much more important in an urban environment. Sidewalks become more than just routes for passage. Opportunities to stop and enjoy a view should be considered, and hidden corners and exposure to high-speed traffic should be avoided. Small-scale textures,

FIGURE 2-21 *For most bridges, the dominant dimension in the environment is horizontal. Note also the strong shadows on the pier and side slopes and the vertical minishadows at the stiffeners over the pier that make them seem larger.*

FIGURE 2-22 *An example of an urban freeway where the surroundings create a vertical emphasis. Even here the horizontal lines of the bridges dominate. I-5, San Diego, California.*

FIGURE 2-23 *Illusions. Visual illusions applicable to bridge design.*

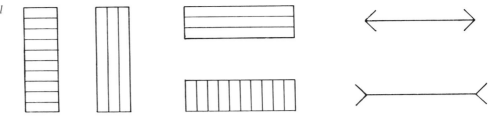

details, and special materials may not be noticed on a freeway, but they can be valued components of an urban structure.

Illusions

Since the viewer's aesthetic reactions to the bridge will be almost completely created by his or her visual perceptions, we must be aware that what people *perceive* is not always what is *there*. The visual sense is susceptible to manipulation and illusion. Illusion can work both to the designer's advantage and to his or her detriment. It is the engineer's job to recognize where the potential for illusion exists and put it to good use. The sketches in Figure 2-23 give examples of common visual illusions that have applications to bridges. (See also Figure 2-24.)

Understanding all Three Dimensions

To accurately analyze the appearance of a bridge, techniques must be used that illustrate what people will actually see at the viewpoints from which they will see it. The standard two-dimensional engineering drawings—plan, section, and elevation—are very deceptive in presenting how the bridge will actually look. The typical two-dimensional elevation drawing is particularly misleading, since only the first columns of the piers appear in the drawing. In reality, all of the columns will be visible from almost every viewpoint. Compare Figure 2-25 with Figure 2-26.

Tools must be used that portray the three-dimensional reality of the structure: models, perspective views, and photographs with the bridge inserted. However, even perspective views and photographs can deceive unless they are taken from an appropriate viewpoint. The typical aerial oblique rendering of a bridge is essentially irrelevant for understanding what a bridge will look like to most observers. That view will only be available to the occasional low-flying helicopter pilot (Figure 2-27).

FIGURE 2-24 *The effect of a visual illusion created by the lines of the abutment appear to "stretch" the girder, making it seem longer and, therefore, thinner. The shadow dividing the girder into two horizontal strips reinforces this effect.*

FIGURE 2-25 *This bridge looks very simple in an elevation drawing. Ingraham Street Bridge over Mission Bay, San Diego, California.*

FIGURE 2-26 *The view of the actual bridge, however, is more complex.*

Drawings must be taken from the viewpoints of the most likely observers of the bridge. However, the helicopter view does have some use in explaining how the bridge fits into its site.

Views of bridges over water should be taken from the most important points along the nearby shore (Figure 2-28). Views of bridges over highways should be taken at driver's eye height, from positions in the traveled lanes of the underpassing roadway, at distances of 300 to 500 feet (see Figure 2-29). Each bridge will have its own set of relevant viewpoints. Not all viewpoints can be covered by photos. The designer may have to extrapolate from one photo to other locations.

For projects involving a series of bridges, a replacement of a bridge that is one of a series of bridges, or a bridge that will be seen from an adjoining highway, it is worth driving through the project or along the adjoining highway to understand what can be seen at the speeds users will be travelling. The experience can be recorded with a video camera and stills from the video used for further analysis.

STRUCTURAL ART

Keys to Success

As shown by the paired examples in Chapter One, the better-looking bridges have the following characteristics in common:

- They are simpler, there are fewer elements.
- They are relatively thinner.
- The lines of the structure are continuous.
- The shapes of the structural members reflect the forces on them.

You will notice that this is a list of characteristics: simplicity, thinness, and continuity. It does not include specific features, like Y-shaped piers or haunched girders.

FIGURE 2-27 *The Sassafrass River Bridge, Maryland. The helicopter pilot's view, an attractive but mostly irrelevant viewpoint.*

FIGURE 2-28 *The Sassafrass River Bridge, Maryland. Most people's view, in other words, the relevant viewpoint.*

That's because most bridges differ from one another in important ways: The site requirements are different; the available technologies are different; the cost environment is different. If the engineer of a new bridge seeks to achieve beauty simply by imitating the features of previous, successful bridges, he or she will usually misuse those features. He or she will apply them in ways that are not appropriate to the bridge at hand. Characteristics, on the other hand, can be applied to any bridge.

Would other bridges with these same qualities also be considered attractive? Yes. A review of the work of outstanding bridge engineers will answer that question. Look at the work of Robert Maillart (Figure 1-18, page 13), Christian Menn (Figure 2-35, page 49), and several state Departments of Transportation (Figures 1-12, page 9; 1-13, page 10 and others), and you will see that these qualities are apparent.

The most-admired modern structures create their visual impression with the forms of the structural members themselves. To take an example from history, think of the Eiffel Tower (Figure 2-30). The shape of the tower is a direct translation of the moments created by the wind forces acting on the tower. The tower's open frame-

FIGURE 2-29 *Visualizations of the driver's experience need to be taken from actual driving-eye height in order to give an accurate impression. This view was developed for the study of the I-74 crossing of the Mississippi River near Davenport, Iowa, described in Chapter Three.*

work of laced iron struts is a clear example of how to minimize the effect of the wind on a high tower. The tower was condemned by the art and architecture establishment of its time, yet it survived to become the symbol of a city and a nation. It was designed by an engineer who had made his reputation designing bridges, bridges that exhibit the same talent for subtly shaping structure to respond to the forces at work.

Philosophical Underpinnings

At this point a question naturally arises: Why do we agree that these particular characteristics make bridges more attractive? We do not *know*, in a factual sense, and therefore discussions of aesthetics necessarily become somewhat speculative. However, in the centuries since the Greeks first codified their ideas on aesthetics, many theories about preferences in the realms of art and architecture have been developed. The theories can be organized into four categories: *geometric, rationalist, sculptural,* and *structural*.

Geometric Theories

These theories are based on the idea that we find certain things or experiences beautiful because they resonate with patterns built into our cognitive systems. It is a common experience from music that certain combinations of sounds (chords) are generally judged to be beautiful, while others are considered dissonant, even uncomfortable, such as the sound of fingernails on a blackboard. Perhaps the fingernail sound causes our cells to vibrate in ways that disturb their functioning.

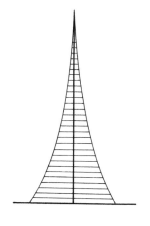

FIGURE 2-30 *Eiffel Tower with diagram of wind-induced moments.*

Many of the classical theories of proportion trace back to this idea. Rectangles proportioned according to the "Golden Mean" (1:1.618) were considered automatically superior because they supposedly appealed to some inherent sense of visual proportion. Classical Greek architecture was the first of many wonderful architectural styles based on mathematically determined proportions. (See Figure 2-31.)

Recent scientific studies have generated some support for this idea in the area of color. Exposure to certain colors is now known to have specific physical and psychological effects. In the bridge world, William Zuk has used statistical samples to determine a commonality of public preference for certain colors on bridges.[9]

Rationalist Theories

This set of ideas states that we like objects whose shapes clearly reflect their function. For example, a teapot with a comfortable handle and efficient spout will be more attractive than a teapot shaped without regard to these functional necessities. Shaker furniture has maintained its appeal despite the demise of the Shakers because it is shaped to respond to its function. (See Figure 2-32.)

This theory applies more easily to objects with a single obvious function. Bridges fall clearly into this category. Buildings accommodating multiple functions are not as easily approached this way.

Sculptural Theories

The goal of sculpture is to produce three-dimensional objects that have the sole purpose of evoking emotions or reflecting ideas. These theories state that we like certain objects because their shapes evoke emotions or reflect ideas that accord with our needs or value systems.

That does not mean that objects with other primary purposes, buildings and bridges, for example, may not also evoke emotion or reflect ideas through their shapes. The Pont Neuf was intended to reflect the dignity and power of Paris as the capital of France.

There is a point at which the sculptor's desire to create shape simply for the sake

FIGURE 2-32 *Shaker boxes use a minimum of material to create elegant and useful containers.*

of its emotional impact conflicts with the economy required of the bridge engineer. A sculpture costing even $100,000 is a far different thing, in terms of the commitment of public resources, from a bridge costing $30,000,000. That is why the best bridge engineers have always embraced the discipline of economy in their work.

In our times, the equality of democratic governance, the efficacy of technology, and the need for economic efficiency are strongly held values. Objects that reflect these ideas will be appreciated.

The engineer's goals of economy and performance coincide well with these values, as long as the engineer finds some way to reflect his or her efficiency in the appearance of the structure (see Figure 2-33). The great engineers have found ways to do that, which is one reason why their works have been judged beautiful by our culture.

Structural Theories

These theories state that we like certain objects because they clearly reflect their structural behavior. They assume that we all possess an inherent understanding of structure based on each person's struggle with gravity from the day we begin to walk.

Consider a cantilever. To most people, a cantilever is more attractive if it is thicker at its support and tapers toward its free end. People find it attractive because it reflects the cantilevers with which they are already familiar: the shapes of their own arms and legs, the branches of a tree. From this familiarity people develop an intuitive understanding of what the engineer knows from calculation: that the stresses in a cantilever are largest at the support, and therefore most of the material should be located there. People will look at a cantilever of constant depth and feel that there is something wasteful about it. (See Figure 2-34.)

Based on that intuitive understanding, people can reason in the other direction as well. When they see a shape thicker at some locations and thinner at others, they assume that it is so because the forces are greatest at the thickest points. They feel that the designer has let them in on the secret: this is how the bridge works. Salginotobel (Figure 1-18, page 13) can be immediately understood even by a layman as two stiff elongated elements resting against each other and against the mountainside, carrying a relativly thin roadway on their backs. It generates an "aha" moment: "Now I know what holds it up." Such moments are the source of much satisfaction for the viewer. They are one reason why cable stayed bridges like the Sunshine Skyway (Figure 1-3, page 20) have captured the public's imagination. Regardless of how difficult they are to design and build, it is easy to see how cable stayed bridges work.

Creating such moments requires the pursuit of what Niehls J. Gimsing calls structural honesty,[10] where the sizes and shapes of structural elements reflect the flow of forces through them. He gives as an example the rules for shaping an honest arch according to the principles of statics:

- The arch shall be curved where it is subject to a distributed load;
- The arch shall have a sharp bend where it is subject to a concentrated force;
- The arch shall be straight where no load is applied.

Christian Menn's Reichenau Bridge (Figure 2-35) is a fine example of these rules. The arch is curved in the central section where it is incorporated in the deck and bearing a uniform load. In the outer sections the arch is sharply bent at the dis-

FIGURE 2-33 *Rogue River Pedestrian Bridge, Grant's Pass, Oregon. Elements shaped primarily for structural efficiency and then refined to bring out their sculptural qualities.*

FIGURE 2-34 *Variations in cantilever shape compared to the moment diagram of a cantilever.*

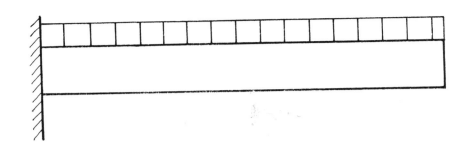

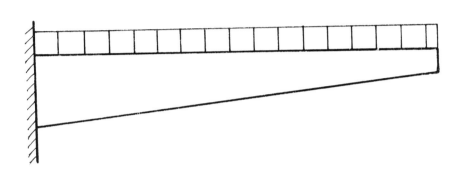

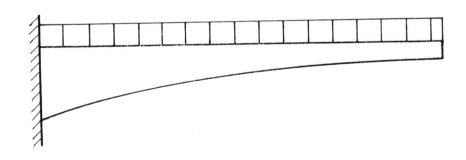

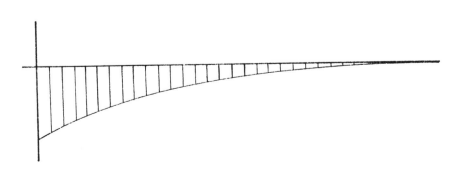

tinct points where slender walls transfer the deck load to the arch and almost straight in between, curving only enough to recognize the uniform load created by the self-weight of the arch rib itself. Because the position of the resisting material is exactly matched to the position of the loads the arch can be very thin, giving the bridge uncommon grace. Contrast this to the many massive arches whose geometry is based on nonstructural curvatures derived from historic architecture.

Study of the above theories is indispensable for anyone who is serious about improving his or her aesthetic abilities. All of these ideas have something to contribute to bridge design. In this area there is much room for debate. Nevertheless, the results of the "exercise" given on page 5 of Chapter One, and the acclaim that bridges like the Salginatobel (Figure 1-18, page 13) have retained through the years can best be explained by structural theory.

Making Structural Art

The potential of bridge design as a means of artistic expression has been recognized by scholars of art and art history as well as engineers. David Billington believes that bridges belong to a medium of their own, "structural art," an art on the same level as architecture and sculpture, but having its own characteristics. Structural art is based on three disciplines:

- *Efficiency:* minimum materials controlled by safety
- *Economy:* minimum costs controlled by serviceability and maintenance
- *Elegance:* maximum personal expression of the designer controlled by the disciplines of efficiency and economy[11]

To put it another way, for bridges *aesthetic excellence lies in the perfection of engineered form,* where perfection has a visual dimension alongside the dimensions of performance and cost. It is important to realize that perfecting the visual dimension must be an acknowledged part of the designer's effort. It will not automatically result from the perfection of the other two. The designer must make the achievement of aesthetic quality a conscious goal and take specific steps to meet it. Efficiency and economy contribute, as Paul Gauvreau puts it, by "generating and refining visual forms that would otherwise not have been created,"[12] but the designer must still con-

FIGURE 2-35 *A structurally honest arch. Christian Menn's Reichenau Bridge, Switzerland.*

sciously shape the result according to his or her ideas of what looks good. Structural art is intrinsically the engineer's art, because only engineers can anticipate the potential of new materials, construction techniques, and design methods to generate new forms, and only engineers can understand how to shape a member for improved appearance without compromising its performance or cost.

We talked in Chapter One about the public's desire to have its bridges take advantage of modern technology, make wise use of resources, achieve economic efficiency, and be responsible to the environment. To the extent that these attributes are evident in the appearance of the bridge, the public will find the bridge attractive.

We also said that people want their cities and towns to be attractive places to live. They want their bridges to be positive features of these places. Bridges are always part of a larger scene. The scene will affect how people perceive the bridge, and the bridge will affect how people perceive the scene. All of this means that the bridges should reflect some awareness of the things around them.

Finally, given the difficulty of the driving task and the paramount need for safety, aesthetic objectives in the highway environment should aim for clarity, balance, simplicity, harmony, and a sense that the environment is satisfying expectations. Elements expressing discord, conflict, and ambiguity are best left to the museums.

These ideas can be generalized into the following principles, which should be the aesthetic goals for any engineer designing a bridge:

- *Simplicity.* There should be a minimum number of different elements; elements doing similar jobs should be similarly shaped.

- *Apparent thinness and transparency.* Elements should appear slender; views through the structure should be preserved (Figure 2–36).

- *Structural clarity.* Elements should be shaped to respond to the structural job they do; what each element does structurally and how it does it should be visible (Figure 2–37).

- *Appropriateness.* The bridge as a whole should have a clear and logical relationship to the things around it (Figure 2–38).

- *Unity with interest.* There should be enough variety to hold people's interest, but all elements should appear to contribute to a single whole (Figure 2–39).

FIGURE 2-36 *Simplicity and carefully shaped "standard" elements result in slenderness and transparency in Tennessee.*

FIGURE 2-37 *This rigid frame makes its structural workings clear: it is thick at the joints, where the moments are the highest, and thin at the ends, where the moments are the lowest.*

FIGURE 2-39 *The consistent angles of the pipe truss and the branching pier give this bridge a sense of unity while the truss and pier have sufficient complexity to engage the viewer's interest. Jörg Schlaich's Nesenbach Valley Bridge, Stuttgart, Germany.*

FIGURE 2-38 *An arch always seems appropriate for a steep, deep gorge. Crooked River Gorge, Terrebonne, Oregon.*

Judgments about appearance can be just as definitive as judgments about structural members, safety, or cost. Merely reducing the number of elements or the numbers of different types of elements can improve the appearance of a bridge. For example, replacing light poles on the outer shoulders with light poles in the median will improve appearance simply by cutting in half the number of elements required for lighting. As we will see, judgments about appearance are based partly on facts (facts about how people see things), and partly on opinions (opinions about what people like about what they see). With regard to appearance, the basis of fact is relatively narrower, and the basis of opinion relatively wider than in other areas of engineering. Therefore, these statements cannot be as prescriptive as those in the other areas. They should be used as guidelines, always with the understanding that there may be valid exceptions arising out of the needs of a particular site or the designer's realization that this is an opportunity to develop something entirely new.

For these reasons the pursuit of aesthetic quality in bridge design cannot be reduced to a rigid set of rules or, worse yet, a computerized "expert" system. New materials, new ideas, new techniques, or the desires of a particular client may call for new approaches to aesthetics, which may violate the old rules. Indeed, if bridge design is a medium of artistic expression, than such violations are to be expected. As Gauvreau notes, one of the characteristics that bridge design shares with the other arts is that "art challenges existing ideas in meaningful ways." Gustave Eiffel's work is a classic example. His reaction to the possibilities of a new material (iron) changed not only ideas about bridge aesthetics, but ideas about architecture, sculpture, and painting as well. The art world was never the same after it saw the Eiffel Tower (Figure 2-40).

The goal of this book is to facilitate the creation of structural art. The ideas herein may point the way, but the achievement must be the designer's own, based on the application of his or her ideas to the situation at hand.

FIGURE 2-40 *Robert DeLaunay's painting of Eiffel Tower: painting responding to structural art.*

CASE STUDY

The Sunniberg Bridge, near Klosters, Switzerland

A recent bridge by Christian Menn, the Sunniberg Bridge (Figure 2-41) near the international ski resort of Klosters, is an excellent example of structural art. The bridge is clearly visible from Klosters. It is 526 meters long, and the longest span is 140 meters. The deck is 9 meters wide curb-to-curb and carries two lanes. The tallest pier rises about 62 meters from the valley floor to the deck. The pylons rise an additional 15 meters above the deck, giving a pylon height to span ratio of 1:10 versus the 1:4 usually found in cable stayed bridges. Edge girders are about 1.07 meters deep, giving a depth-to-span ration of about 1:136. These departures from the usual proportions of a cable-stayed bridge were specifically directed toward an explicit design intention/vision. The citizens of Klosters asked that the bridge be thin and transparent in order to have as little visual impact on their valley as possible. Menn began from that point. The selection of the basic elements all stemmed from that request. (See Figure 2-42.)

Pylons sized as for a typical cable stayed bridge would have projected 35 meters above the deck. The tallest pylon would have been 97 meters high. This would have brought the pylon tops roughly level with the windows of Klosters, which lies at the head of the valley. The short pylons stay well below this level. When viewed from Klosters they are often hidden by intervening vegetation.

This solution of the aesthetic problem created some structural challenges. With a pylon height to span ratio of 1:10, the forces in the cables increased significantly. Typical cable stayed pylons are relatively thin and therefore flexible. Unbalanced cable loadings under unbalanced live loads combined with thin pylons would have created significant pylon deflections and therefore significant girder deflections. In response, Menn made the pylon thicker at the top to stiffen it against longitudinal deflections (Figure 2-43).

Because the bridge is curved, the bridge deck can respond to temperature changes by expanding and contracting radially, carrying the piers along with it. This eliminates the need for expansion joints at the abutments. With no need for expansion joints, the deck can be anchored at its ends. The deck can thus stabilize the pylon/piers longitudinally against deflection due to unbalanced live loads and laterally against wind and other transverse loads. The additional axial forces due to the low cable angles and the longitudinal stabilization of the towers puts relatively large axial forces into the deck and its edge girders, which were strengthened near the pylons, where the axial forces are the greatest, to guard against buckling.

The pier/pylon must respond to a number of forces. The longitudinal restraint created by the deck and the footings creates longitudinal moments which decrease to a minimum at about one-third pier height and

FIGURE 2-41 *Christian Menn's Sunniberg Bridge near Klosters, Switzerland (see color insert).*

FIGURE 2-42 *The townspeople wanted a bridge that was thin and transparent, with as little visual impact on the valley as possible.*

then slightly increase again as the pier nears the ground. Menn shaped the pier to respond to these moments. In the longitudinal direction the piers are thinnest at a point about one-third of their height above ground and flare outward above and below that (Figure 2-44).

In the transverse direction, additional moments are created in the pier/pylons by the lateral movement of the deck and by the eccentricities between the points at which the cables attach to the deck and pylons

FIGURE 2-43 *The statical system for the bridge can be determined by viewing the shapes of the structural members.*

(caused by the curvature of the bridge). The piers could not just be made solid. Some flexibility is required to allow them to move laterally as the deck expands and contracts. So Menn used a vertical vierendeel truss (Figure 2-45.1). The horizontal struts of the truss also stiffen the pier legs against buckling. The pylons are flared at the top to keep the cable stays clear of the curved roadway edges (Figure 2-44). Menn smoothly continues this flare into the pier legs below the deck, bringing the pier legs together so that they are half as far apart at the bottom as they are at the top, further reducing the forces in the pier legs caused by the lateral restraint of the deck. The pier legs themselves are hat-shaped in plan, giving them maximum stiffness against local buckling with a minimum of material, and creating deep shadow lines up the pier legs that make them look even thinner than they are. The result is an elegantly shaped transparent pier that allows views in all directions.

In the overall view, the thin deck seems to float above the trees, cradled by the towers. There are no embellishments, unless you call the pattern of construction joints on the piers an embellishment. All of the features that create the aesthetic impression arise from the shapes and sizes of the structural members themselves, and the shapes and sizes of the structural members arise from engineering considerations. A viewer can understand how the bridge works by studying the shapes of all the major elements. Menn states the goal clearly in his book *Prestressed Concrete Bridges:* "The visual expression of efficient structural function is a fundamental criterion of elegance in bridge design."[13]

In Figure 2-41 the stays connecting to the tower on the left are much more visible than the stays for the remaining towers. At the time the photo was taken, a debate was underway concerning the color that the stays should be painted. The stays on the left are white, the others are black. The white stays stand out against the wooded background, while the black stays blend in. Based on the overall desire to make a bridge with the minimum visual impact on the valley, the final decision was to paint all of the stays black.

The cost of the Sunniberg bridge was 20 million Swiss Francs in 1998.

FIGURE 2-44 *The flared pier tops allow the stays to clear the roadway on a curve (see color insert).*

It was about 15% greater than that of the cheapest of the other bridge designs proposed for the site. The increase amounted to about 0.5 percent of the cost of the entire Klosters Bypass project. The canton engineer and the people of Klosters apparently thought that this additional money was well spent. Not only did it preserve one of their major assets, their scenic appeal, but it added another, "a magnificent monument to their tradition of bridge engineering."[14]

The results of Menn's bridge design process may be summarized as follows: The structural members themselves, shaped in response to engineering considerations, to both illustrate how they are functioning and create a memorable aesthetic impact. Success depends in part on the shapes as required by the forces involved. But through his or her choice of structural type and relative sizes, an engineer can steer those forces where he or she wants them to go, thereby developing a shape that meets his or her aesthetic vision. Once the basic shapes are determined by engineering considerations, the exact curve of the flare, the exact proportions of the cross struts and other features can be refined to achieve visual elegance (Figure 2-45.1).

Menn calls this step "artistic shaping." In this endeavor, engineers are seeking to tap the aesthetic pleasure that can be created by an attractive shape. The sculptor Constantin Brancusi also sought to tap the appeal of pure shape in pieces like his *Bird in Space* (Figure 2-45.2). The difference is that the engineer's shape must start from the requirements of his or her structure.

Banzinger Bacchetta, Partner of Chur, did the detailed design for the Sunniberg Bridge, Switzerland, based on Menn's conceptual design. More details of the analysis and design can be found in an article by Chelsea Honigman and David P. Billington in the May/June 2003 issue of the *Journal of Bridge Engineering.*[15] More on Menn's overall approach to bridge design and his other works can be found in *The Art of Bridge Design, A Swiss Legacy.*[16]

FIGURE 2-45.1 *Refined elegance in a structural shape based on structural role and size (see color insert).*

FIGURE 2-45.2 *Refined elegance in a sculptural shape by Constantin Brancusi (see color insert).*

THE TEN DETERMINANTS OF APPEARANCE

How people react to a bridge depends on what they see first, followed by whatever additional features they notice, in the order that they notice them. First impressions are rarely completely overridden by later information. People see the big elements first, the structural elements. Thus, the shapes of the structural elements are the most important in determining our reaction to a bridge. Color of the big elements is next; then, if time and distance permit, the details and their colors and textures. It follows that the 10 determinants of a bridge's appearance are ordered in importance by their influence on the shapes of the structural elements:

1. *Vertical and horizontal geometry.* Determines how high the structure is, whether it is curved in one or two planes, how the roadway widens and splits, how it is positioned relative to prominent surrounding features.

    Before there is a concept for a bridge, the geometry itself establishes a ribbon in space that will be attractive or unattractive. The geometry establishes the basic lines of the structure, to which all else must react. A graceful geometry will go a long way toward guaranteeing a successful bridge, while an awkward or kinked geometry will be very difficult to overcome with later decisions.

    Bridge engineers are often handed the geometry as a predetermined element. They must reserve the right to evaluate it and request changes if necessary to improve the appearance of the structure.

2. *Superstructure type.* Defines whether the structure is an arch, girder, rigid frame, truss, or cable-supported.

    By establishing the overall shape of the structural members, this decision establishes the most memorable aspect of the structure.

3. *Pier placement.* Establishes not only the points at which the structure contacts the topography but also the size of the openings framed by piers and superstructure.

    The success of the visual relationship between the structure and its surrounding topography will depend heavily on the apparent logic of the pier placement. For example, a pier near a water edge will appear more logical if placed on the shore side of the boundary. People understand that it is easier to build a pier on land than in the water, and crossing the water is why the bridge is there in the first place.

    The openings between the piers have a shape that will influence the impression a bridge makes. Placing the piers to create well-proportioned openings will improve the appearance of the structure.

4. *Abutment placement and height.* Determines how the bridge starts and ends and, for shorter bridges, how the structure is framed.

FIGURE 2-46 *Geometry, the starting point and base for all that follows.*

FIGURE 2-47 *At a glance, the shape of the superstructure establishes the viewer's first impression. Gunpowder River Bridge, Joppatowne, Maryland.*

The abutment placement also establishes the shape of the end–span opening, which can have a significant influence on what can be seen beyond the structure, and how well the structure relates to adjoining uses.

5. *Superstructure shape.* Establishes the form of the structural members, including deck overhangs, parapets, and railings.

   This is the point at which the structure can be shaped to respond to the forces on it. The intrinsic interest of the structure will be determined by this characteristic.

6. *Pier shape.* Defines the form and details of the piers.

   From many viewpoints, particularly at oblique angles to the structure, the shapes of the piers will be a major influence on the impression created.

7. *Abutment shape.* Defines the form and details of the abutments.

   For shorter structures, and from viewpoints near the ends of longer structures, the shape and detail of the abutment will be important. For structures involving pedestrians, the provisions made for them at the ends of the bridge can be among the most memorable aspects of the structure.

8. *Color.* The colors of the uncoated structural materials as well as the coated elements and the details.

FIGURE 2-48 *Pier placement establishes a logical relationship of the Glade Creek Bridge to the West Virginia topography by taking advantage of natural promontories on the sides of the valleys.*

FIGURE 2-49 *These minimal abutments allow the space of the highway and the North Dakota countryside to flow through the bridge.*

FIGURE 2-50 *The appeal of this haunched girder bridge in Idaho owes much to the shape of the superstructure and its blending with the piers. Bennet Bridge, Idaho.*

FIGURE 2-51 *These memorable piers are shaped to reflect their function.*

Color, or lack thereof, will influence the effect of all the decisions that have gone before. It provides an economical opportunity to add an additional level of interest.

9. *Other bridge details, surface textures, and ornamentation.* Establishes elements that can add interest and emphasis.

Structural elements, such as stiffeners and bearings, can serve this function. Indeed, traditional systems of architectural ornament started from a desire to visually emphasize points where force is transferred, such as from beam to column through an ornamental capital. Patterns of grooves or insets and similar details added to make an element seem thinner are other examples.

10. *Signing, lighting, and landscaping.* Though not actually part of the structural system, these elements can have great influence on the aesthetic impression a bridge makes.

Decisions about the first five determinants are usually thought of as strictly "engineering" decisions. However, they are inescapably aesthetic decisions as well. Decisions on determinants 6 through 9 are the ones most often thought about when speaking of "bridge aesthetics," but it is almost impossible for decisions made about these elements to completely compensate for poor decisions made about the

FIGURE 2-52 *The abutments of the bridge shown in Figure 1-8 (page 4) not only help frame the Rocky Mountain view, they give an indication of their role in supporting the girders.*

FIGURE 2-53 *The bright yellow color of the steel box girders make this interchange a memorable milepost for travelers using I-440 in Nashville. This outstanding design for a complex interchange is made possible by the torsional stiffness of box girders. (See also color insert.)*

FIGURE 2-54 *Structural details, such as this combination of horizontal and vertical stiffeners, can serve a traditional ornamental role by emphasizing points-of-force transfer.*

FIGURE 2-55 *Signing, lighting, landscaping and the bridge itself can be integrated into an attractive whole. I-5, Olympia, Washington.*

FIGURE 2-56 *All 10 Determinants of Appearance make a positive contribution to this structure.*

first five. A poorly shaped girder cannot be corrected by painting it an attractive color. Make the best possible decision about the first five elements, and then use elements 6 through 9 to accentuate and improve the positive qualities that have been created.

The tenth determinant lists elements that are often added to the structure or placed next to it. Depending on their size, placement, and detailing, they can be irrelevant, positive, or very negative contributors to the appearance of the bridge. In most highway situations the same organization controls all three, thus leaving no excuse for a careless and unattractive result.

IMPROVING AESTHETIC SKILLS

The best way of improving aesthetic skills is by careful observations of the bridges seen every day. Look at existing bridges. Go back to the same bridge at different times of day and at different seasons. See how the changing sun angle and changing light intensity affects what you see. Take a camera and photograph the bridges. Take a sketch pad, soft pencil, and/or Magic Marker and draw the bridges. Don't worry if the drawings are not very good. You don't have to show them to anybody else. The important thing is to look at a bridge with a fresh eye, to see what it actually looks like, not what you assume it looks like from your experience of other bridges. The act of photographing or drawing will help you do that (Figure 2-57).

FIGURE 2-57 *By studying existing bridges, we can discover the qualities that make them seem to fit into their environments. CA 1 over Pescadero Creek, California.*

Compare your observations. Do you like the appearance of one bridge more than the next? Why? Which feature appeals to you? Why does it have that positive effect? Does it add a shadow line, change the texture of the wall, make the girder look thinner? Keep notes on your photos and/or drawings, and organize them to use in your own designs.

Color is a subject that can best be learned through personal observation as well. But because bridges in general have such a limited range of color, it is necessary to look elsewhere to experience a wider range of the possibilities. Our appreciation of color is influenced by size and exposure, so that buildings furnish the nearest approximation of the effect of various colors. Look at the colors of buildings, and make judgments about what you see. Bridges are usually seen against a largely natural background, so the effect of background colors must be absorbed as well. Comparative observations and color photography are indispensable tools in this process.

It is a process that can be formalized through case studies, familiar features of law and business schools. The case study summarizes the important facts about a given bridge in a few pages. These facts include a description of the major physical characteristics of the bridge, its structural capacity and safety features, construction and maintenance costs, and appearance. The study should also include an analysis of the success of the bridge and the presentation of alternative features that would improve future bridges of the same type. Figures 1-23.1 (page 22) and 1-23.2 (page 22) show an example of such an analysis. The key is an *integrated* study of each bridge, one that equally considers performance, cost, and appearance. Considering appearance without considering cost is just as irresponsible as considering cost without considering appearance.

Training yourself to think in three dimensions (3D) is an important skill. It encourages you to visualize, from the beginning, what the bridge will look like in real life. (It helps in understanding the performance aspects of bridge design as well.) As you design, try drawing elements in three dimensions even from the roughest beginning sketches. Your field observation sketches will give you practice. Learn how to use the 3D module of your CAD program. The better ones are easy to use if the models are kept simple. There is more on 3D thinking in Chapter 3.

Many successful bridge designers have put their ideas about bridge appearance on paper. Some of these publications include specific and practical guidelines. Fritz Leonhardt's *Brucken*[17] is an example of such a book that includes many specific guidelines distilled from the author's experience. Christian Menn's *Prestressed Concrete Bridges*[18] similarly includes an eloquent chapter on his approach to aesthetics in bridge design. The Transportation Research Board has published *Bridge Aesthetics Around the World,*[19] an international anthology of articles by 21 authors from 16 countries. It includes a comprehensive bibliography of everything written in English on the subject in the previous years. David Billington's *The Tower and the Bridge,*[20] and his books on Robert Maillart,[21] are also outstanding. For general exploration of the design process, Rudolph Arnheim's *Visual Thinking*[22] is worthwhile. Books like these are useful sources of guidance and inspiration, and should be part of the library of any practicing bridge designer. Details and further suggestions can be found in the notes following each chapter.

1   Gauldie, Sinclair, 1969. *The Appreciation of the Arts 1, Architecture.* London: Oxford University Press.

2   This section relies in part on an unpublished 2003 paper by Martin P. Burke, Jr. titled *Bridge Aesthetics: A Discipline in Need of a More Effective Language.*

3   Most of the definitions in this and the following section were originally developed by the author for the Ohio Department of Transportation's *Aesthetic Design Guidelines,* 2000, Columbus, Ohio

7   Tunnard, Christopher, and Pushkarev, Boris, 1963. *Man-Made America, Chaos or Control?* New Haven, CT: Yale University Press.

9   Zuk, William, 1974. "Public Response to Bridge Colors," *Transportation Research Record.* Washington, DC: Transportation Research Board, vol. 507.

10 Gimsing, Niehls J., 1999, *Bridge Aesthetics and Structural Honesty,* International Association for Bridge and Structural Engineering Symposium, Rio de Janeiro.

11 Billington, David P., 1983. *The Tower and the Bridge: The New Art of Structural Engineering.* New York: Basic Books, Inc.

12 Gauvreau, Paul, 2003, *Three Myths of Bridge Aesthetics,* University of Toronto, Toronto

13 Menn, Christian, 1990. *Prestressed Concrete Bridges,* 2nd edition, ed. and trans. P. Gauvreau. Berlin.

14 Billington, D.P., 2003, *The Art of Bridge Design: A Swiss Legacy,* Princeton University Art Museum, Princeton, New Jersey., p. 192

15 Honigmann, C. and Billington, D.P., 2003, Conceptual Design for the Sunniberg Bridge, *Journal of Bridge Engineering,* American Society of Civil Engineers, Volume 8/Number 3, Reston, Virginia

16 Billington, *Art of Bridge Design,* 2003.

17 Leonhardt, Fritz, 1982. *Brucken.* Cambridge, MA: The MIT Press.

18 Menn, 1990.

19 Transportation Research Board, 1991. *Bridge Aesthetics Around the World.* Washington DC.

20 Billington, *The Tower and the Bridge,* 2003.

21 See Acknowledgements, page 13, and the endnotes for Chapter One, page 25, for more books by Billington.

22 Arnheim, Rudolph, 1969. *Visual Thinking,* Berkeley, CA: University of California Press.

chapter *three*

DESIGNING A BRIDGE: PRACTICAL PROCEDURES

"Beauty will not unconsciously arise out of a search for economy. Rather, there are choices for the engineer to make, and he is to be judged on them."

—DAVID P. BILLINGTON, AMERICAN ENGINEER AND EDUCATOR

We have now covered all of the basic ideas and concepts. The moment of truth is at hand. The engineer is asked to design a bridge. A blank piece of paper waits on the drawing board. How to begin?

THE BRIDGE DESIGN PROCESS

Many engineers see the bridge design process as the analysis of forces on a predetermined structural system and the sizing and detailing of structural members to resist those forces. In fact, that effort is only the third of the four parts of the bridge design process. The first part is the development of a *design intention* or *vision,* which defines all the things the bridge is intended to accomplish and all of the goals that it is intended to meet. The second part is *conceptual engineering,* which investigates a wide range of structural possibilities to define the one that will best meet the design intention/ vision. Then comes *analysis and detailing.* The fourth part is the *construction* stage, which must be monitored to ensure that the design is followed and the design intention/ vision satisfied.

The Design Intention/Vision

Before a designer can start on the bridge itself, he or she must have an idea of all of the criteria that the structure must meet and all of the concerns that will act on the structure. Many bridge engineers focus on the physical and technical concerns such as navigational clearances, pavement widths, and utility loads. In recent years, the Federal Highway Administration and many other transportation agencies have recognized that this should be a much broader task. They are promoting Context-Sensitive Design/Context-Sensitive Solutions as techniques to incorporate all of the concerns

FIGURE 3-1 *The engineer's starting point.*

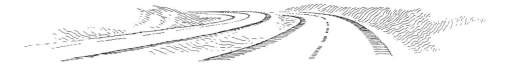

that should act on a project. These techniques provide a framework to recognize important but less technical concerns such as environmental resources, community plans and aspirations, and urban design goals (Figure 3.2). The final result of this process is the integration of all of these matters into a written statement of design intent, or vision.

Bridges being public works, this is inevitably a collaborative process. At the very least, the owner (or the owner's representatives) will set minimum physical criteria and budget. Financing and review agencies will also be involved in a significant way. It is not uncommon, for example, for environmental agencies, the Coast Guard, or the Army Corps of Engineers to have a major say in span arrangements and bridge length. If the bridge is prominent in a community, or has historical or symbolic associations, community groups and historical societies will want to play a role. This is the point at which to include the community and elected officials. All of these parties must be identified and invited into the collaboration process at this beginning stage, before any direction has been set. By involving all concerned parties from the beginning, the designer can have some confidence that the final result will address the most strongly held desires of the community and meet with approval at later stages of design. If this step is neglected or abbreviated, the consequences will almost certainly include delay and added cost when the design must be revised to accommodate previously ignored views.

All parties should be considered part of the project team, jointly charged with developing the best possible bridge for the site. This will ensure that every participant starts off with the same understanding of the project, and that all will feel like valued contributors. Indeed, the insights of nonengineering members of the team concerning the nontechnical issues will be particularly valuable at this stage. Later sections of this chapter will discuss the elements that must be considered in developing a design intention/vision and techniques to make the collaborative process as productive as possible.

FIGURE 3-2 *The underside of this German bridge provides an attractive and visually interesting ceiling, shelters the light rail stop, reflects daylight into the underbridge area, and supports the transit wire.*

For bridges covered by the National Environmental Policy Act, the identification of the design intent/vision is often begun during preparation of the environmental impact statement (EIS). Visual studies of potential bridge alternatives can also make their own contribution to EIS preparation. The development of the Design Intentional vision is discussed later this chapter.

Conceptual Engineering

Concept Development

Once the design intention has been established, conceptual engineering can begin. Conceptual engineering is the stage when all plausible options and some not-so-plausible options are examined in sketch form. The end product of this stage is the formal Type, Size, and Location presentation. Engineers consider this the point at which 30 percent of the design is complete. This phase is often called Preliminary Design. However, the traditional Preliminary Design process often jumps too quickly to the final concept, based on assumptions about what has worked previously in similar situations, as was discussed in Chapter One. There are many possible solutions to a given problem. The more that are considered, the better the chance that a unique solution will be found that best fits the design intention/vision.

The Swiss engineer Christian Menn describes this stage as follows:

> During the development of a bridge concept, the structural idea becomes solidified through scaled diagrams of the major elements and, parallel to this, through just enough calculations to demonstrate feasibility and economy. This conceptual design stage requires the designer to develop a four-fold vision of the project, simultaneously considering issues of structural form, mathematical analysis, construction methods, and the relationship of the structure to the site.[1]

Conceptual engineering begins with the quick generation of a wide range of possibilities, going beyond the usual and accepted options, with a quick development of at least the first seven Determinants of Appearance. For each option, the possibilities for geometry improvements should be investigated first, then structural types,

FIGURE 3-3 *Flaming Geyser Bridge, designed by Kevin Dusenberry/ENTRANCO, Bellevue, Washington. Restricted underclearance and the bridge's function as the entrance to a county park led to an unconventional design that answered both functional and symbolic requirements.*

pier locations, shapes of the major members, and so on, through pier and abutment shape. Color, railings, and surface treatment—the more detailed Determinants of Appearance—can wait until later. In order to make this process as clear as possible, the balance of the book is organized around the Determinants of Appearance, presented in their order of importance.

Starting with a wide range of seemingly unusual options is critical. By definition, improvements will come from new ideas, ideas beyond the currently understood range of possibilities. By starting with a wide range of options, the designer is less likely to miss that key idea that will lead to an improved structure and, perhaps, to a great bridge.

Promising structural concepts should be sketched in three dimensions to get a sense of how they will appear in finished form. Rough calculations, sufficient to set major member sizes and get preliminary costs, should be done. The next step is an equally quick evaluation, discarding the alternatives least likely to achieve the design intention/vision. Next, the more promising alternatives are refined, including refined 3D visualizations. Then the review and evaluation process should occur again, always looking at structural effectiveness, cost, and appearance at the same time. At each successive stage, the drawings should be more accurate and more carefully rendered.

There will be dead-ends and blind alleys. Sometimes the process will have to be recycled, if, for example, a decision to change structural type indicates that a change in vertical alignment is warranted. Through the generation and review of multiple alternatives, the process will narrow down to a basic concept that meets all three criteria: efficiency, economy, and elegance. Figure 3-4 shows an example of the result of this process.

We can imagine how this process might go for a project crossing a valley. Part of the design intention may be a desire to maintain open views up and down the valley. In response to this concern, the designer might start off with a set of alternatives that relies on large spans to keep the views open. Figure 3-5 shows a solution of this type. The side effect of long spans, of course, is that the structural members themselves become quite large, creating in themselves an impediment to the views.

FIGURE 3-4 *By taking a new look at what can be done with precast I-girders, the designer came up with a haunched pier section that allowed a longer span and kept the piers out of the water. High Bridge replacement, Snohomish County, Washington.*

FIGURE 3-5 *Canyon Creek Bridge, Anchorage, Alaska. Long spans require larger members, which themselves can begin to curtail the view.*

The designer might try a different approach: alternatives with very short spans supported by a multitude of very thin members, as shown in Figure 3-6. The views are still available, but now they are seen through a transparent lacework of thin structural members. The aesthetic impression of the second set of alternatives is obviously quite different, as are the structural and cost implications. The shorter spans actually

FIGURE 3-6 *Portage Viaduct over the Genesee River, New York, is an example of the transparency possible with thin structural members.*

present some possible economies. By considering these along with the appropriate visualizations, a decision can be made about which concept best achieves the design intention/vision

Testing Concepts in Three Dimensions

Quick, three-dimensional sketches are the best way to try out multiple ideas at the early stages of design development. The sketches in Figure 3-7.1 were done by Art Elliot when he was chief bridge engineer for the California Department of Transportation. They helped to resolve a dispute with nearby residents over the piers for the San Mateo Bridge (Figure 3-7.2). *Design Drawing,* by William Kirby Lockard[2] offers easy-to-learn and easy-to-use techniques for quick 3D sketching. Three-dimensional sketches are indispensable when developing your ideas in the early stages of design.

Modern computer-aided design (CAD) systems offer the most important method of easily generating three-dimensional views of alternate designs from the important viewpoints. Three-dimensional (3D) CAD models make it possible to

FIGURE 3-7.1 *Art Elliot's sketches of possible pier designs for the San Mateo Bridge.*

FIGURE 3-7.2 *The San Mateo Bridge as built.*

achieve a level of accuracy not easily attainable with hand sketches, particularly for bridges with horizontal or vertical curvature. Accuracy is important, as small differences in girder depth or pier width can make enormous differences in the final appearance of the structure. Having access to such 3D representations during the early design stage is so valuable that the CAD operator should be made an integral part of the design team from the very beginning of the conceptual engineering stage. With the help of a skilled CAD operator, the designer can develop the concept first in three dimensions, then reduce it to two dimensions only when it is necessary to begin the construction drawings.

In a few hours, a basic 3D model of the deck, piers, and girders can be developed in the computer and placed on generic backgrounds of earth and sky. The model can be rotated for viewing from different locations. Alternative superstructures and substructures can be quickly inserted and evaluated. The designer can try out different designs and proportions, understand exactly how they will look, and quickly make adjustments, before spending much time on details and calculations. Most CAD programs move seamlessly from 3D to 2D drawings. As one example, the 3D module for MicroStation uses the same geometrical and design information that must be entered into the computer to prepare 2D drawings. Thus, it is just as easy to enter it as a 3D model first. Once the 3D model has been refined, negligible additional time is required to create a 2D drawings from it. In sum, there is no longer any excuse for not using 3D views in bridge design. This is too valuable a tool to be left until the formal presentation at the end of the process.

In preparing such drawings, all visible elements, such as signing, lighting, and guardrails, should be included, at least in generic form. Conversely, all elements that will be below-ground or hidden should *not* be shown. If visible, they will distort the aesthetic analysis.

The CAD drawings shown in Figure 3-8 were developed during the early stages of design for the new NJ 52 crossing of Great Egg Harbor Bay in New Jersey, to study two different pier options. The views were taken from the Dockside Marina and Café, a waterfront gathering place in Ocean City. From the drawings it became clear that the Y-shaped pier made the bridge seem more massive and less transparent (option A).

As additional features of the bridge are developed, they can be added to the 3D computer model. Perspective drawings from the CAD system can then be superimposed on photographs taken from the relevant viewpoints. Because of the accuracy of CAD, the bridge can be made to match the photograph. A series of such rendered CAD drawings over photographs is the most accurate method of presenting alternative bridge designs. Because of the sense of familiarity created by the photographic backgrounds, these drawings are particularly valuable at public meetings: The background is complete and easy to recognize, enabling viewers to get a good idea of the relative size of the bridge by comparing the drawing to familiar features of the existing scene.

Creating such drawings has specific technical requirements. Each CAD drawing must be done from the same viewpoint and eye height as the camera that produced the base photograph. The photographer should avoid wide-angle or telephoto lenses, as these will introduce distortions into the photograph. The rendering of the CAD drawing to create photorealism requires a CAD operator with an artist's sensibility

FIGURE 3-8 *This CAD study of two pier shape options for the New Jersey 52 Causeway Replacement, taken from a popular local restaurant, allowed the designers and the public to make realistic visual judgments about structural elements at an early stage in the design.*

OPTION A

OPTION B

OCEAN CITY BRIDGE OPTIONS
ROUTE 52   OCEAN CITY, NEW JERSEYS
Scale: NTS   5-20-03

ROUTE 52 CAUSEWAY REPLACEMENT PROJECT
Michael Baker Jr., Inc.
Rosales Gottemoeller and Associates, Inc.

for shadows and reflections. Figure 3-9, made for a proposal for the new St. Croix River Bridge, Stillwater, Minnesota, illustrates the degree of photorealism that can be achieved by a skilled operator.

Such drawings do, however, have two major technical limitations. For accuracy's sake, they are limited to a field of view no wider than 60 degrees, which is the human eye's range for recognizing detail and judging distance. Thus, the photo cannot show the effect of peripheral vision, or of the wider field of view humans unconsciously take in by small horizontal eye movements. The result seems like an unreasonably limited slice of the actual structure. To address this limitation, the temptation is to use wide-angle lenses for the photo and computer simulation in order to show more of the structure. Unfortunately, this distorts the result by making the structure seem farther away from the observer than it really is. This limitation can be better addressed by using cameras that take panoramic views of the background. Then visualizations can be prepared that combine the panoramic view with the computer model. Such views can also be animated. The "camera" is made to pan over a long structure as an actual observer would turn his or her head to take it all in.

This technique still falls short of showing the effect of movement around a structure. To do that requires 3D modeling of the actual background in the com-

FIGURE 3-9 *Computer-drawn visualization of a proposal for the new St. Croix River Bridge, Stillwater, Minnesota, showing how it would look from a nearby marina.*

puter. Unfortunately, at the current state of animation art, entering and manipulating realistic and detailed topographic information demands unreasonable computational power and time. To compensate, the backgrounds are simplified, giving the product a cartoon-like quality. However, such presentations can still be useful for judging size and shape in relation to surroundings, and in analyzing proportions. They are particularly useful for analyzing the driver's experience (Figure 2-29 page 44), which is, as noted previously, much like a movie. Again, the emphasis should be on real eye heights and likely paths for actual observers. The computer can put the "camera" anywhere, from lying underground to swooping overhead. Designers can be seduced by this capability into creating images that no one would ever see in the real bridge.

In spite of all of the realism made possible by the combination of 3D computer renderings and photographs, there is still a place for a well-done artist's watercolor or colored-pencil drawing. CAD-based visualizations rarely capture the "feel" of the final product as successfully as an artist's rendering. Perhaps it comes down to the people shown in the drawing. Somehow, the artists' versions always seem more animated and real. Thus, artists' renderings should always be considered as the final touch for an important public presentation.

Physical models are also valuable tools. They have the advantage of allowing the viewer to move around the structure and select any viewpoint he or she chooses. Rough study models can be used in the early stages of design to understand the relationships of complex structures or groups of structures such as interchanges. For public presentations, models are unequaled for generating public understanding of and enthusiasm for a design. Unfortunately, models have several major disadvantages. It is difficult for an observer to put himself or herself at a scale eye height. The view he or she is getting is the helicopter view, which few people will share after the structure is built. The answer to this is to place the model on a raised support, so that the observer's eye height is closer to the scale eye height. Models are also unwieldy to present to large numbers of people. They work best in "open house" meetings, during which people are allowed to individually move around and observe the model. Finally, if there are multiple alternatives, separate models or inserts must be constructed for each alternative.

Keeping the Community Involved

It is necessary, and an integral part of the Context-Sensitive Design/Context-Sensitive Solutions process, to keep the public involved throughout this effort. In this way, designers will have some confidence that their interpretation of the public's ideas is correct, improving the likelihood that the final result will earn a positive reception. Often, seeing the conceptual drawings will call to the public's mind concerns not voiced earlier. Better to get these on the table at this early stage than to have to come back and deal with them during final design.

The 3D visualizations and models prepared as part of conceptual engineering make it easy to productively involve the public. They can understand what is being proposed and comment on it constructively.

Analysis and Detailing

Calculations and Construction Documents

It is only after the basic concept has been selected that detailed calculations and engineering drawings should be started. This is also the point at which secondary elements such as railings, colors, signing, and lighting, elements lower on the list of Determinants of Appearance, can be developed consistent with the selected structural concept.

Such details are important to the final aesthetic success of the structure. The development and selection of details must remain true to the intentions of the original concept. Throughout the final design process, details should be drawn in three dimensions to ensure that they contribute to the design concept. Significant changes should be reviewed to ensure that the design intention has not been compromised.

The 3D visualizations and models prepared earlier should continue to be refined, adding the details such as railings and colors, as they are developed. There is no better way to test a railing design than to show how it will look to a pedestrian on the sidewalk or to a driver trying to see through it to the scene beyond (Figure 3-10). For particularly intricate railing designs, a full-scale mockup of a section can be very helpful.

This part of the bridge design process presents particular hazards if there is a change in personnel between the conceptual engineering/preliminary design and the analysis and detailing/final design stages. Large bridge-building organizations sometimes organize their staff into preliminary and final design groups, or perform the preliminary design in-house and leave the final design to a consultant. There is great danger that something significant will be lost in the transition from one group to the next. The preliminary designers must incorporate a continuing review function throughout the final design process, to ensure that all changes are positive and that all community commitments are met.

FIGURE 3-10 *A computer-modeled pedestrian's view of a railing design for the Clearwater Memorial Causeway replacement described in Chapter Six.*

Selection and Standardization of Details

Standard details have an important place in bridge design. They are especially important when they represent the distillation of hard-won functional experience, as in a crash-tested railing. Indeed, many agencies attempt to design their bridges by agglomerating standard details. When some detail has proven to be attractive or useful on one bridge, it is particularly tempting to apply it to all subsequent bridges.

Any aesthetic success experienced through these approaches will be sheer coincidence. All bridges are unique. Each deserves a fresh look. The functional aspects of standards will always apply. The visual aspects need to be reconsidered for each bridge. Some will apply completely, some will apply partially and have to be modified, and some will not apply at all, and new details will have to be developed to fit the specific situation. The New Mexico bridge illustrated in Figure 2-15 (page 34) demonstrates a good use of standard details, in this case developed specifically for the Big I interchange.

An important part of the role of the engineer is to carefully review the standard details that will apply to his or her bridge and suggest an alternative wherever the existing standard would compromise his or her design intention. An important function of bridge-building agencies is to periodically take a hard look at their standard details and discard or improve those that are preventing the agency from producing good-looking bridges. See Chapter Seven for more on this subject.

Be Wary of Value Engineering

The typical value engineering process starts with a group of outside experts sitting down with the project staff and brainstorming the basic "values" that the project is intended to meet. The problem is that this group will probably not include a representative of the community or someone prepared to champion the community's values, nor is it likely to include someone who understands the value of the aesthetic features. These values then become easy targets in the cost-cutting that follows, resulting in a bridge that disappoints the community and betrays its trust.

The first step in avoiding these problems is to make the original design intention/vision statement the value statement on which the analysis proceeds. The second step is to make sure that the value engineering team includes representatives from the key stakeholders and people who understand aesthetic values. If this group then agrees on a less expensive way of meeting some community or aesthetic goal, or decides that some community or aesthetic goal is too costly, fine. At least all involved will have had a chance to participate in that decision and will understand the outcome.

Continue to Keep the Community Involved

The Context-Sensitive Design/Context-Sensitive Solutions process requires keeping the community involved right through design. Some community members will be more interested in the choice of railing designs, colors, and other details decided in this stage than in the structure itself. As a practical matter, community involvement is most productive in the early portions of this stage, when alternative designs for railings and other details are being considered. The community will not have much to contribute later when the major issues are weld sizes and rebar patterns.

Construction

Construction inevitably brings to the surface problems that were inadequately recognized during the design or generates new ideas by the construction team itself. These matters need to be addressed in ways that preserve and enhance the design intention/vision. Doing so requires that the original community stakeholders and design team continue to be involved throughout the construction process.

DEVELOPING A DESIGN INTENTION/VISION

The remainder of this chapter talks about the design intention/vision phase. Chapters Four and Five provide information to support Conceptual Engineering, Analysis and Detailing, and Construction. Chapter Six describes how all of this comes together in actual bridges.

Understanding the Site and Its Community

Identify the Physical Requirements

The engineer must begin by gaining a full understanding of the requirements and the site. There will be the usual engineering requirements: horizontal and vertical alignment, required clearances, stream flows, foundation conditions, traffic, and loads. These will determine the range of possible structural types.

Explore the Visual Environment

The next step is to fully understand the visual environment. Here a camera is indispensable. Visit the site. Walk around it. Find the places from which the bridge will be visible. Identify and prioritize the major viewpoints. Take pictures—in color. Lots of them. Get the important ones enlarged.

If the bridge is part of a larger project in which the bridge's environment has not yet been built, then the engineer must use more imagination. The highway drawings should be used to generate three-dimensional views from the likely viewpoints. These can be placed on photos of the larger background to get a sense of what the site will look like.

At this point, it is also necessary to get a sense of whether the night view will be important, and in what way. Finally, the differences imposed by the seasonal changes need to be understood.

Determine Nearby and Associated Uses of the Surroundings

The engineer must consider the surrounding area and its uses. Will the bridge adjoin the town's Little League field? Do people picnic or boat or bicycle near the bridge? Under it? Will any of these uses change as a result of new development or new zoning? Is the town trying to improve its attractiveness to industry or tourism, and should that affect the bridge? The presence of these activities will, of course, affect the selection of viewpoints and the priorities among them. It will also affect the criteria that might be established; for example, the appearance of the underside of the structure might gain importance, depending on the use of the area below the bridge.

In any situation where there will be pedestrian usage under the bridge, or for high bridges where the underside can be seen from many viewpoints, the designer should consider the appearance of the underside. Such situations often occur over urban streets and in parks. In these cases, the designer is creating an outdoor room with the ceiling being the underside of the bridge, as shown earlier in Figure 3-2 (page 62).

Environmental requirements should be determined at this point. For example, the presence of rare marine life may restrict the options for pier placement; or restrictions on construction seasons caused by the presence of rare birds may influence the choice of structural type.

Identify Symbolic Functions

Does the bridge have a symbolic function? What are the desires and aspirations of the people who will see or use the bridge? A prominent structure, such as a bridge near the center of a town, will become a landmark that will come to symbolize the town. The people of the town will have strong feelings about what this bridge should look like, and those feelings need to be understood. Even less prominent bridges will have neighbors and users who will want to see a bridge that enhances their neighborhood and their daily travel. It is important to understand what it is about their neighborhood or their trip that they consider important.

Establish the Boundaries of the Design

The boundaries of the design should be based on all elements that will be seen simultaneously from the major viewpoints. All elements connected visually need to be visualized and developed together: bridge, retaining walls, noise walls, guardrails, signing, lighting, traffic signals, and landscaping. If the first person to see everything together is the contractor, then it will be surprising if it looks good.

Writing the Design Intention/Vision Statement

The designer must next decide which of the factors he or she has learned will influence the design, and in what priority. All of the preceding information should be organized and evaluated for its effect on the structure: identify the viewpoints with the highest priority; evaluate environmental and land use information; identify critical aspects and features in the surroundings. The product should be a list of potential design considerations for the structure. This list of factors, both "hard" and "soft," needs to be considered, prioritized, and integrated into a statement of the goals that the bridge is intended to address. The community, review agencies, and all relevant stakeholders need to be involved in this process.

As example goals, the presence of a nearby civic center may indicate benefits to be gained, in terms of views and space for crowd passage, from longer-than-normal spans, while the presence of crowds under the structure may add a desire to have an attractive underside. In other locations, the color or form of a nearby rock outcropping or buildings may be established as an important consideration (Figure 3-11). The preservation of significant existing views may be placed as a constraint on pier location or type. The statement of intent/vision must then be adopted or confirmed by the consensus of all of the participants in the process.

For the project shown in Figure 3-11, the New Jersey 52 Causeway Replacement over Great Egg Harbor Bay, a formal Community Partnering Team was organized, combining community leaders, interested citizens, review agencies, the New Jersey Department of Transportation, and its consultants. The group developed a list of guiding principles for the project, which included:

1. Make the new bridge people-friendly, accommodating marine operations, pedestrian/bicycle use, fishing, and access to the islands, as well as environmental protection.
2. Make the new bridge a visual asset and enhancement to Great Egg Harbor Bay and both towns.
3. Reopen the sweeping views of the bay from the shore and the water by making the main channel portions of the new bridge as transparent as possible by the use of long spans and thin piers.

FIGURE 3-11 *When the time came to replace the NJ 52 bridge in Somers Point, a community partnering team was formed to create the design intention/vision.*

4. Open up the bridge users' views to the bay and its shore by the use of see-through railings so that users can enjoy the panorama of the natural environment and the adjoining towns.

5. At both ends of the bridge, establish gateway areas that preserve and enhance the historic resources of each community and open up views to the historic resources so that travelers are more aware of them.

6. Create an aesthetic design that is consistent across the entire bridge, compatible with both communities, and complementary to their historic elements.

7. Light the bridge and its roadway so that the project is an enhancement and an asset to the bay and the two towns at night, as well as by day.

8. Accommodate an Ocean City Visitors Center and Scenic Overlook with due consideration to the historic elements of the neighboring communities and the environmental sensitivity of the area.

These principles, along with the requirements for navigational clearances, maintenance of traffic, and other technical issues, became the design intention/vision that guided the design of the bridge. Figure 3-8 (page 68) shows an early study of bridge options aimed at meeting this design intention/vision.

Frequent Considerations in the Development of a Design Intention

Corridor and Urban Design Themes

Major highways often incorporate a number of bridges seen in close succession or even at the same time. Concern for the quality of the sequential experience requires that the appearances of all of the bridges be considered together. Concept themes should be considered on all new and reconstructed routes during the earliest stage of design. This is particularly important when some of the bridges are to be done by different designers or different agencies. A design theme can be developed by selecting a common vocabulary of bridge elements and applying them more or less consistently to all of the bridges on a given route or in a given area. For example, a standardized parapet profile can be developed and used consistently throughout a series of bridges.

Standardized colors, certain surface materials, or a standardized texturing for retaining walls and abutment walls are other obvious devices to develop a theme. A concept/theme does not require that all structures be identical. Variations along a route can be used to influence the user's frame of mind, such as gradually reducing spans as the highway moves into an urban area. The theme becomes one of change or difference. But the change or difference must be controlled and compatible over the entire route.

Themes may not be appropriate for all corridors or all structures within a corridor. The exception to the rule makes its own statement. However, the designer should remember that the absence of a theme is in itself a theme, with its own set of perceptions and reactions.

The major challenge of a theme is to reconcile the common features of the theme with the need for each bridge to address its particular structural requirements. If all of the bridges share similar structural aspects, the problem is obviously simplified. However, where a wide variety of structural situations exist, it becomes a more

FIGURE 3-12.1 *The difficulty of finding a compatible theme to meet many different situations. This pier looks fine when it is short.*

FIGURE 3-12.2 *However, when the pier is tall, it looks bottom-heavy.*

difficult challenge to find a theme that allows each structure to develop its own efficiency, economy, and elegance while still contributing to the larger ensemble. In such cases, reliance on a structural element (for instance, a standardized pier design), may produce disappointing results when the element gets stretched to meet all of the different situations (Figure 3-12.1 and 3-12.2). Better in those cases to rely on nonstructural elements (parapet profile, color, surface texture) to carry the theme.

Structures in interchanges must be given special consideration, as several of them are usually visible at one time, and they may even be physically linked. This requires not only application of the common elements resulting from a design theme, but also close coordination by the designers of each structure.

The introduction of a new crossing on a route with an existing theme creates a real challenge in order to establish a positive relationship to what exists. It re-quires a thorough evaluation of the location, noting existing structures near the new, their pier placement, and shapes, color, and details. This does not mean that the new must match the existing. A choice has to be made: Should the new structure attempt to complement the existing structures, or should it make its own statement?

If all of the existing structures are similar and very close together (300 feet or less), it is probably better to closely complement the nearby structures. Attempt to tie the new to the old with common features, details, or color. For example, a two-span continuous crossing in an area with four-span bridges could be tied to the older bridges through a similar pier design or parapet profile. When all of a series of closely spaced structures are different, it is probably better to select the best one and attempt to complement it, rather than insert a new variant into the visual cacophony. If the existing bridges are more widely spaced, greater flexibility exists. The engineer can look to criteria outside the freeway itself, such as the desire to emphasize the presence of a nearby town, as reasons to complement or not complement existing bridges.

Historical or Contemporary Architectural Styles as Themes

On occasion, a bridge will be required to respond to a theme established to encourage a certain design flavor in a larger area of which the bridge is a part. These themes are often expressed as variations of some historical or contemporary architectural style. The legitimacy and success of such efforts, with regard to the bridge, depends on the nature and specificity of the design requirements. If the requirements set general guidelines for the use of certain materials, colors, or details, they can usually be accommodated within the engineer's disciplines of efficiency, economy, and elegance. However, if they establish specific requirements that compromise the goals of efficiency, economy, and elegance for the bridge itself, then they should be strongly resisted.

Let's examine the difficulties with historic styles first. Historic architectural styles were based on the technologies and materials available in their times: brick, stone, cast iron, and wood. Traditional architectural details are suitable for those materials and reflect the aesthetic and social values of those times. Today's designers, in contrast, have available high-strength steel and post-tensioned concrete. It is not unusual for them to build multispan structures as long as a city block with spans of more than 200 feet that could cover a dozen traditional buildings.

There is another problem in trying to adapt traditional architectural styles to bridges. All traditional architectural styles are based on reference planes that are straight and level. Bridges are rarely straight or level. It is impossible to make a curved highway bridge on a 5 percent grade and as long as a whole block of colonial buildings look like Georgian architecture. The attempt will result in something that is neither a good bridge nor good architecture—and will be very expensive as well.

Figure 3-13 shows a set of structures built in the monumental center of St. Paul, Minnesota, near the capital building. Rather than placing the ashlar stone of the capital building on the bridge superstructure, the designer closely imitated the stone with precast concrete, and then used it at locations in piers and retaining walls consistent with the historic use of stone and with consistent architectural detail. This is very successful in the central part of the area, where the highway is straight, and the crossing structures are close to level and cross at right angles, so that the neoclassical architectural details have the rectilinear reference planes for which they were origi-

FIGURE 3-13 *Capital Center Structures, I-35E, St. Paul, Minnesota.*

nally designed. However, even this system breaks down where the highway curves away from the downtown, and the cross streets slope across the highway on significant vertical grades (Figure 3-14). Then the traditional architectural details look unstable and out of place.

One frequent requirement imposed by an historical theme is the use of a traditional material, such as brick. Such materials can be easily and sensibly accommodated when used in their traditional fashion as walls and piers (Figure 3-15.1). However, if the requirement extends to mandating the use of brick to face the bridge girder itself, then substantial cost and durability problems will be inflicted on the bridge without producing an offsetting improvement in aesthetic quality (Figure 3-15.2). Most people will find unsettling the sight of brick suspended in midair over spans of 150 feet.

As pointed out in Chapter One, when communities ask for a historical theme they are often saying what they *don't* want. They don't want a featureless girder bridge that looks like 99 percent of the contemporary bridges they know. They *do* want a bridge with the memorability and interest of a historic bridge. Designers should show the community representatives examples of contemporary bridges with

FIGURE 3-15.1 *Appropriate use of traditional materials on walls and piers: U.S. 50 over U.S. 301, Queenstown, Maryland.*

FIGURE 3-15.2 *Using brick to face a bridge girder looks unsettling; the heavy masonry seems to have no visible means of support.*

equal levels of memorability and interest. Figures 3-17, 3-18, and 3-19 in the next section are examples. As part of the discussion, the public should be made aware of what their bridges will say about their community. A bridge displaying historical architectural details is saying that its community lives in the past, that its best ideas are behind it. Once made aware that there are contemporary alternatives that will give them the desired level of memorability and interest, the community may well decide on a contemporary structure. The best approach is always to build a modern bridge that is as true and as high-quality a reflection of the current era as the historic buildings are of the past. In the process, the community will mark itself as one that is looking confidently into the future.

A theme based on a contemporary architectural style may avoid some of these problems as long as masonry cladding on structural members is not required. The designer should remember that details and materials suitable for pedestrian exposure on buildings will be wasted on a bridge. They will be unseen and unappreciated by viewers traveling at highway speeds.

Replacement of a Historic Bridge

The prospect of replacing a well-loved historic bridge often mobilizes people whose first preference is to simply reproduce the existing bridge. However, there are many potentially appropriate options, ranging from a memorable contemporary structure with little resemblance to the original bridge to a contemporary structure that reflects the original bridge in form and size to a close replica of the bridge to be replaced, with endless gradations along the spectrum in between. Hard-and-fast guidelines cannot be given. Much depends on the community's goals for the area and, indeed, its view of itself.

The Naval Academy Bridge shown in Figure 1-22 (page 19) was built as a result of an international design competition. The bridge is in Annapolis, Maryland, a colonial town that still serves as the state capital. The bridge itself adjoins the historic buildings of the U.S. Naval Academy. The competition jury, which included representatives of the nearby neighborhoods and the local historical society, decided that the best bridge would be one that embodied all the modern virtues of simplicity and transparency. Their reasoning began with the high quality of the existing scene. They felt that a new simple and transparent bridge would preserve the focus on the historic buildings, rather than become itself the centerpiece.

Notwithtanding their much longer history, Europeans seem to have an easier time with this notion. The concrete box girder shown in Figure 3-16 carries a major street across the upper reaches of the Rhine River into the medieval center of Stein am Rhein, a Swiss tourist and resort town. Apparently, the citizens thought the medieval town had sufficient architectural quality to hold its own, without assistance from "historic" detail on the new bridge. Indeed, the contrast with a contemporary bridge makes the medieval buildings even more impressive. Paris is sometimes given as an example of a city with many historic bridges. However, each bridge was built using the best engineering and material available in its own time, and the entire ensemble nevertheless retains its charm. The newest vehicular bridge across the Seine, the Pont Charles de Gaulle, opened in 1993, is a streamlined structure adjoining structures built 100 years before with all of the architectural detail characteristic of their era (Figure 3.17).

FIGURE 3-16 *The people of Stein am Rhein, Switzerland, felt that the medieval buildings of their town center had enough architectural strength to hold their own and did not require the addition of medieval decoration to the town's new concrete box girder bridge.*

FIGURE 3-17 *Pont Charles de Gaulle in Paris adjoins bridges built 100 years earlier.*

An intermediate option is to build a bridge that emulates the form of the existing bridge, using, for example, the same span arrangement and structural system, while incorporating modern materials and techniques and omitting the historical detail (Figure 3-18).

These options ought to be considered first, through the process described previously. If, after seeing and considering contemporary examples of comparable memorability and interest, the community still feels that only a new bridge that re-creates the traditional appearance of the existing bridge is acceptable, then the goal must be to do it correctly, so that the new bridge truly emulates the style it is aiming for. In these cases, it is usually not possible or desirable to reproduce exactly the existing structure. New structural techniques, new materials, and new functional require-

FIGURE 3-18 *The Lake Redding Bridge in Redding, California, emulates the spans and structural system of its existing bridge neighbor, but remains clearly a bridge of its own time.*

FIGURE 3-19 *The High/Main Street Bridge in Hamilton, Ohio, emulates the historic arch that it replaced and includes traditional details consistent with the adjoining War Memorial on the right.*

FIGURE 3-20 *The Third Street Bridge in Napa, California, provides all of the elements of a traditional bridge but with a thin deck and contemporary shapes that make it clearly a bridge of our time.*

ments will make that an unreasonable solution. Effort should instead be directed at looking for the memorable features in the existing structure and finding ways to reinterpret them in modern technologies. For example, say the existing bridge is a deteriorating earth-filled concrete arch. Perhaps a precast concrete arch could be built, with longer spans suitable to the modern material, but with the same general shape. Or, perhaps, there are details of railings—or lighting—or traditional materials on the old bridge that can be incorporated in a new structure while the major structural members themselves are shaped to take advantage of new materials. The High/Main Street Bridge now under construction in Hamilton, Ohio, is an example of this approach (Figure 3-19).[3] At the insistence of the local community the bridge was designed to emulate the historic earth-filled arch that it replaces. The structure is built of modern precast concrete girders shaped to resemble the old arches while details were developed that are consistent with the adjoining War Memorial.

Building a New Bridge Paralleling an Existing Bridge

The first determination when building a new bridge that parallels an existing bridge should be the life left in the existing bridge and the period of time the two are expected to exist together as an ensemble. If the old bridge has a limited life and/or is likely to be replaced in a different form for other reasons, such as navigational clearances, then there is little reason to be concerned about differences in appearance. However, if the opposite is true, the appearance of the existing bridge must be taken into account. The goal is not necessarily to make the two bridges identical, but to make them appear complementary as an ensemble. Figure 3-18 (page 80) shows a good example of this approach. If the bridges are in an historic area, much of the analysis in the preceding section will also apply.

Begin by analyzing the old bridge. If the old bridge has good qualities, then these should be carried over to the new bridge (and preserved on the old bridge). The desirable features can be made the basis of common details, common proportions, and common structural systems. For example, it is usually desirable, and often necessary for hydraulic reasons, to make the pier placements the same, or at least make the spans of the new bridge some multiple of the spans of the old. This analysis can be carried all of the way into details such as railings and pier facings. Again, the new features need not be literal reproductions of the old, but should have some correspondence in shape, rhythm (such as in railing post spacing), and scale. For example, it may not be necessary to construct new fieldstone abutments to match old abutments: horizontal rustication in the new abutments with the same spacing as the old stone coursing could tie the two together.

Color offers a quick and obvious way of establishing correspondence between old and new. If a new steel bridge is to be built next to an old one, it is a simple matter to paint them the same color or a related color. Concrete may offer similar possibilities. For example, if the older pier cap is to be given an epoxy finish for maintenance purposes, the same finish can be applied to a similarly shaped new pier cap adjacent.

Reconstruction and Rehabilitation

It is also possible to reconstruct old bridges to incorporate new requirements and materials while respecting their original design intent. When older bridges are rebuilt or rehabilitated, the question becomes how much, if any, of the old detail should be restored? This question can be answered by determining the answers to two others: Of what visual quality is the detail, and how important is it to the historical record and/or the community's image of the bridge?

If the old detail is of very high quality and/or is very important to the community, then every effort should be made to restore it to its original quality or to incorporate this detail in new forms, using the new materials used in reconstructing the bridge. If not, the reconstruction should seek to simplify the appearance of the bridge in ways that bring out whatever positive qualities exist in its structural members and overall form.

INVOLVING COMMUNITIES

Community involvement is an integral part of the Context-Sensitive Design/ Context-Sensitive Solutions process, and indeed any responsible public works design process. It must start with the development of the design intention/vision and continue right through construction and even maintenance. Bridges have long attracted detailed citizen interest beyond the usual not-in-my-back-yard reaction. For example, the Rainbow Bridge (Figure 3-21) over the American River has long been considered the defining landmark of Folsom, California. Images of the bridge appear in the masthead of the local newspaper and in the logos of several shops on the main street. When the need arose to add an additional crossing nearby, the community was adamant about being intimately involved with the design of the new structure. The community wanted to be sure that the new structure would be of equal visual quality and that its appearance would complement the old bridge. Until recently, it has been difficult for engineers to respond to such concerns.

Appearance is in itself a difficult subject to deal with in public works. Facts do not apply to appearance to the same degree as other aspects of public works. The arena deals largely with opinions. Everyone is a taxpayer, so each believes his or her opinion is as good as anyone else's, deserving to be heard equally.

FIGURE 3-21 *Rainbow Bridge, Folsom, California.*

A complicating fact is that most people, including engineers, have difficulty imagining the appearance of the final structure based on engineering drawings. That makes agreement almost impossible. After all, how can people agree on, or even discuss, appearance when each person has a different mental image of the object?

In the past, such problems could not be easily solved because of the difficulty and cost in presenting alternatives in ways that could be understood by the general public. Perspective drawings were expensive to prepare, were subject to artistic license, which reduced their credibility, and were limited to fixed viewpoints. Models are also expensive and give a false impression of size. Now, with the advent of computer-aided design (CAD), as discussed earlier, it is possible to economically prepare realistic depictions from multiple viewpoints. Citizen groups can easily understand what the bridge will look like, and thus become constructively engaged in judgments about the alternatives.

There is, of course, an element of risk in involving the community. Members may jump too quickly to an obvious or comfortable choice and miss a better option. Or they may fall in love with a design that turns out to be too expensive or impractical, leading to unsatisfactory compromises. The community may decide to build something fashionable that is hard to maintain or that they soon come to dislike. In the 1950s and 1960s, automobile companies relied on momentary enthusiasms for catchy new features, such as tail-fins on Cadillacs, that people would quickly tire of, leading them to go out and buy a new car. That is not an option open to buyers of bridges. As the knowledgeable professional of the process, it is the engineer's job to anticipate these possibilities and lend his or her expertise and, yes, prestige, to educate the public and guide the process. Doing so requires an appreciation of public relations, group dynamics, and politics. The key is to recognize that a productive community involvement process takes time. Trying to build consensus too quickly risks an unhappy result.

Foundation of Positive Citizen Participation

The first rule of citizen participation is to *listen*. That point cannot be overemphasized. The whole purpose of the citizen participation effort is to understand what citizens think about *their* design problem. There is only one way to do that: listen to what they have to say. It is not the easiest thing to do. Most citizens start a citizen participation process by letting off steam. They have some real or imagined dissatisfaction with the project or the agency and they want to "get it off their chests." Or they have some pet idea about the project, which, likewise, they have to get off their chests. These statements can become quite emotional in their delivery.

The natural tendency for an agency representative or consultant is to take the emotion personally and become defensive. The representative tries to explain or defend previous decisions or to explain why some pet idea will not work. These reactions are mistakes. No useful exchange will take place until the attending citizens have voiced their concerns. Until that happens, they aren't prepared emotionally to *hear* anything that the consultants or agency representatives might say. Indeed, an attempt to respond at this point will only confirm a frequent suspicion that the whole process is a sham and that the agency is not really interested in listening to the community. The only way to get over these hurdles is to listen. Once the citizens recognize that the consultants and agency representatives are indeed interested in what they have to say, and understand what they are saying, the emotion will dissipate and they will be prepared to begin a useful dialogue.

The second rule of successful citizen participation is to avoid expert overload. As the project proceeds, there is a risk that citizens will become intimidated by the professional knowledge of consultants and agency representatives and not volunteer anything of substance. Of course, this result makes the citizen participation process useless. For example, citizens often say things or make suggestions that have obvious technical flaws. There is a tendency on the part of the professionals to immediately step in and correct the technical part of the statement. This reaction is also a mistake. It only reinforces the perception of a knowledge gap between the citizen and the professionals. The better reaction is simply to let the matter pass and continue to listen to the rest of the message, which will often indicate areas of citizen concern or perception about their project of which the technical person is unaware. If the technical flaw becomes substantive as the discussion proceeds, the professional can find a way to correct it at a later time, without seeming to contradict, or "show up," the citizen's lack of technical knowledge. Again, the best approach is to listen.

The use of technical jargon is a related danger. Everything an engineer has to say can be said in plain language. One of the enjoyable challenges of citizen participation is rediscovering the plain-language equivalents for all of the things engineers deal with. If the process continues successfully, the professional will find the citizens beginning to understand the technical concepts, even with the jargon.

Indeed, the best definition of a citizen participation process is a *mutual educational process* between the engineers and the citizens. The engineers are educating the citizens about appropriate bridge types, costs, other technical issues, and, most of all, the ability of modern bridge design to respond to the citizens' aspirations; the citizens are educating the engineers about community concerns, perceptions, and values for the site in question. Like any educational process, it must begin with the willingness on the part of the *educatees* to listen to and understand the *educators*. The best way for both groups to get to this point is through face-to-face dialogue and the resulting familiarity that will evolve.

Once the substantive dialogue begins, the agency and the citizens will find they have different but overlapping goals. The agency's goal will be to make citizens more aware of the real possibilities, constraints, and costs governing the engineering decisions, so they will be able to understand and accept a reasonable, cost-effective solution. The citizens' goal is for the engineers to come to understand the community's concerns and aspirations, so that the engineers will be able to develop a feasible solution that responds to these concerns and to produce a structure that meets the community's needs. In the best projects, the citizen participation acts as a spur to the engineers, unleashing their creativity and producing a proposal that is improved beyond what would arise out of a conventional engineering process.

Useful Techniques

There are seven guidelines for a successful participation process:

1. *Tailor the process to the project.* The possibilities here range from an appointed citizen review committee to an open-to-everybody process. The best answer depends on the nature and number of the groups involved, the size and scope of the project itself, and the nature of the political body or bodies that make the final decisions. The prime criterion should be to set up the process so that it meets the needs of these decision makers. For the Folsom project, the mayor appointed a Citizens' Review Committee, with members representing all the interested groups and agencies, including several that opposed the bridge. The lines of communication between engineer and community should extend to the ultimate decision makers. It is important that the decision makers themselves understand the design problems and opportunities sufficiently to make a good decision.

2. *Create opportunities for mutual education by the citizens and the engineers.* The two groups need opportunities to develop a face-to-face relationship and discuss general topics of mutual interest before being faced with the stress of decision making on the project itself. A very valuable technique is a joint field visit to the site of the project (Figure 3-22). The citizens will be on familiar ground. Indeed, they will know more about some aspects of the site than the engineers do. This puts them on a more equal footing with the engineers and makes them more receptive to comments from the engineers about the structural possibilities at the site.

3. *Encourage citizen participation in developing the design intention/vision statement.* The design intention/vision statement should include the community's goals for the bridge. These should be stated in terms of characteristics such as opening and preserving views, or what the bridge will look like from an important viewpoint. This will allow for the consideration of a wide range of structural types that might achieve these goals and help head off a premature decision. Some citizens will come to the process with a general idea or vision as to what they think the bridge should look like. It is very important for the engineers to understand what these visions are. They may in fact be impractical; however,

FIGURE 3-22 *Folsom Citizens'*
Review Committee on field visit
to the new bridge site.

such ideas can be addressed in general terms in the first round of alternative evaluation. Initially, the goal should be to get the citizens to think through and state what they are trying to achieve with their suggestions, and to incorporate that in the design intention/vision statement.

Including the citizens in the development of the design intention/vision statement from the beginning increases the likelihood that they will possess a sense of ownership about the final result, avoiding the need for a later stop and restart. Of course, it is always possible that the citizens will come up with something that the engineers had not imagined. This possibility has to be recognized and encouraged.

At Folsom, as a way of soliciting the citizens' thoughts about the site, a line drawing of the site was abstracted from a photograph (Figure 3-23), with the profile of the new structure added. The members of the Citizens' Review Com-

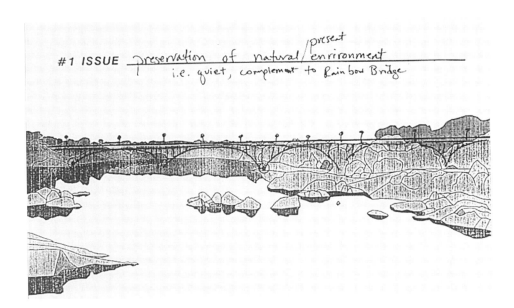

FIGURE 3-23 *One citizen's*
vision for the new American
River Bridge.

mittee were asked to sketch on the drawing their visions of the structure. Arch bridges clearly predominated in the results, but other possibilities also appeared.

4. *Develop multiple alternatives with technical and cost implications and provide several iterations of review and selection.* As explained, the conceptual engineering selection process should start with a wide range of alternatives, at least four, to give citizens a full range of the possibilities, to provide for further education of the citizens and the engineers as to their mutual areas of concern, and, most importantly, to ensure that no promising alternative is overlooked. This is the time for the engineer to be creative, developing alternatives that meet the community's aspirations and the other parts of the design intention/vision while mobilizing the potential of modern materials and design techniques. At the same time, the designer must guard against proposing, in the enthusiasm of the moment, an alternative that will exceed the budget or not meet technical requirements. That will invariably be the alternative that the community falls in love with, leading to unhappy compromises later on.

Even the initial round of alternatives should include rough technical evaluation and costs. If the citizens are to understand the decision-making process, they need access to all of the information. The major engineering constraints and cost implications must be explained. Citizens are taxpayers: Faced with realistic and credible information about cost, they will, in most cases, draw the logical conclusion. However, their logic may not be the same as the engineer's. Citizens may consider it logical, in view of the importance they attach to the bridge, to spend some of their money on features added for their contribution to the appearance of the structure, and they have a perfect right to do so.

Such decisions are often complicated by policy-level considerations about the responsibilities of the various levels of government for the various features of the bridge. For example, the state transportation department might be willing to pay for the major features of the bridge, but not for a nonstandard ornamental railing. These issues can be anticipated in advance, and should be sorted out among the governmental decision makers before the community involvement process begins. The community should have a clear picture from the beginning regarding from which pocket various costs will come and what the budget limitations of the project are.

The first round of review is the time to address alternatives suggested by the community. If impractical, whether by cost or structure, or unattractive, the problem can be addressed before too much time or money has been spent. If all of the citizen visions are not addressed at the beginning, their sponsors will not be able to let go of them. They will continue to raise their ideas, and the whole process may be disrupted or delayed. If their ideas are addressed at an early stage, the sponsors will be able to move on to more practical possibilities.

Some ideas or concerns may not be aroused until a proposal reaches the table. The process has to allow for incorporation of these ideas, with revision, extension, and reevaluation of the possibilities in a second or even third round of review. Multiple rounds of review also allow for the deferral of details to the later rounds of review. There is no need for the basic decisions about structural type to get hung up at the beginning because of debates about railing details.

Finally, the cost estimates need to be accurate. Once a decision is made based on them, it will be very hard to undo if they turn out to be wrong, because the whole community will be involved. Agencies and consultants will find that the usual assumptions about the time and labor required for preliminary design are too low. Sufficient resources must be provided to develop the necessary detail and accuracy over multiple alternatives.

5. *Present all alternatives in three-dimensional form, showing the bridge in its actual setting.* Even engineers have difficulty visualizing what a bridge will actually look like using engineering drawings (Figure 3-24.1 to Figure 3-24.4). It is almost impossible for nonprofessionals to do so. If the attempt is made, the results will be, at best, misleading impressions, and, at worst, endless arguments over differences in individual impressions, none of which are accurate. The available alternatives to resolve this problem, such as three-dimensional CAD-based drawings over photographs, have already been discussed.

6. *Present the result of the process to the responsible elected or appointed body for ratification and approval.* Even with the best citizen participation process, there may remain a disgruntled minority who feel that their views have not been appropriately heard or reflected in the final decision. In order to forestall the later derailment of the process by this group, the selected alternative should be a given a formal review process, during which the formal commitment of the appropriate decision-making body or elected officials is received. This will make it more difficult and less likely for the decision makers to change their minds at a later date should other objections surface.

The Folsom project followed these steps through to the end. The Citizens' Review Committee recommended a three-span arch bridge to the city council based on the information, including cost estimates, developed during the study; and the council ratified their choice. At that point, an unhappy member of the committee resigned and led a petition drive for a referendum on the council's decision. The referendum was opposed by the remaining members of the committee and the council, and it failed. By that time, the whole city was invested in the arch concept.

Unfortunately, it was then discovered that the early cost estimates had not accounted for the seismic effects of the arch on the foundations. The only option that

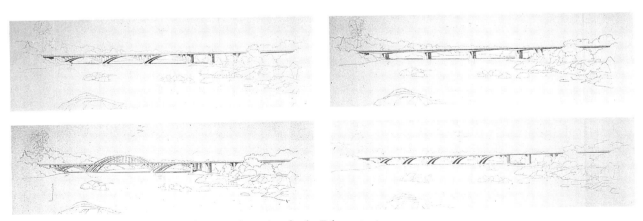

FIGURE 3-24.1 THROUGH 3-24.4 *Preliminary alternatives for the Folsom structure.*

FIGURE 3-25 *The new American River Bridge at Folsom.*

would meet the budget was a haunched girder, because it could take advantage of seismic isolation bearings to reduce foundation costs. This put the council in an impossible position: either find more money or give up on the arches the city had its heart set on. Against the advice of the design team and even the contractor they decided to build a hybrid, a haunched girder with decorative arches (Figure 3-25). The community is happy, but most engineers find it illogical and unattractive.

The community process described in Chapter Six for the Clearwater Memorial Causeway avoided this problem by identifying the additional cost of the preferred alternative at the time it was presented to the city council. They were able to make their decision with all of the facts in front of them, and appropriated the additional funds at the same time that they picked the structure type (Figure 3-26).

A community participation process conducted along these lines makes possible reasoned debate based on a common vision of what the new bridge should look like.

FIGURE 3-26 *The final design for the new Clearwater Memorial Causeway in Clearwater, Florida, met the community's goal to reopen the view to Clearwater Beach, blocked by the existing bridge it will replace.*

While time-consuming in terms of staff/consultant time, the process may well save both staff/consultant time and calendar time by resolving issues in an orderly and timely manner, rather than through an endless series of contentious hearings and redesigns. Best of all, the process will result in a clear and specific mandate for the appearance of the new bridge.

INVOLVING OTHER VISUAL PROFESSIONALS

Chapter One pointed out that engineers must accept concern for appearance as an integral part of bridge engineering. The best bridge engineers have seen their bridges accepted as icons of art as well as of engineering. Gifted engineers working without the assistance of architects, artists, or other visual professionals have produced masterpieces. Thus, it is not necessary for all bridge design teams to include visual professionals. However, with the increasing interest of the public at large in the appearance of their bridges, it will be necessary for all bridges to have a design engineer who is knowledgeable about aesthetic quality.

The engineer's first goal should be to develop his or her skills to meet this requirement. However, for reasons of time or personal inclination, this is not always possible. Accordingly, engineers have often sought the advice of other visual professionals—experts in aesthetics who are consulted in the same way as experts in soils, traffic, and wind. There are many memorable bridges that illustrate the potential success of this approach if done well; the Golden Gate is just one example. Such collaboration does not relieve the engineer of the responsibility to be knowledgeable about aesthetics. As leader of the design team, he or she remains responsible for the final result. Many over-decorated and expensive failures have been created because the involvement was arranged poorly or because someone other than the engineer took over the lead role. In developing this relationship, the engineer should remember the basic difference between engineers and other visual professionals.

The engineer must establish that the visual professional's role is as aesthetics advisor and critic, making comments and suggestions for the engineer's consideration. These suggestions can come from four groups: urban designers, landscape architects, architects, and artists. Urban designers and landscape architects are skilled at identifying the visual and physical relationships between the bridge and nearby buildings and open spaces. Their ideas can often be valuable in suggesting parameters and design features that will improve the fit between the bridge and the community. Examples might include aligning an abutment sidewalk to better serve pedestrians from a nearby residential area, or providing a median opening between structures carrying dual roadways to allow light penetration and a wider range of uses to an area below the bridge. Architects can make helpful suggestions on the proportions and the shapes of major elements, and, particularly, in the design of railings, walkways, visual details, and materials encountered at the street and pedestrian scale. Artists can make suggestions about incorporating works of art into the project. Most visual professionals have the skill to visualize concepts and quickly sketch ideas, which can make it easier to reach the best solutions. Color and lighting can be complex issues requiring specialized technical knowledge and refined visual sensibility. Each field has developed as a consulting specialty of its own. Even trained visual professionals get assistance in these areas; engineers can, too.

If their involvement is to be successful, the engineer must be sure that his or her advisors understand the basic issues involved in bridge design. Most visual professionals are used to dealing with buildings and their immediate surroundings; but bridges are significantly different from buildings. Appearance is a matter of perception, and the perceptions of people in and around buildings are different from the perceptions of people around bridges. People in and around buildings are walking, standing, sitting, or even lying down. Most people viewing bridges are moving at 30 to 70 miles per hour and view the bridge through the windows of an automobile. Their perceptions are significantly altered in ways that are not immediately obvious, as was explained in Chapter Two. Bridges are also larger than all but the largest buildings; thus they are seen from greater distances and have a greater impact on the landscape and the people around them. Small elements that are important at the scale of a building, such as bricks, can become visually lost when applied to a bridge. Bridge loads, too, are generally larger and more dynamic, imposing significantly different requirements on a bridge structure than on a building structure. The thermal and weather exposure of a bridge is more rigorous than that of a building structure, which usually is protected by the building's skin. Most important, as Paul Gauvreau puts it:

> Architects (and other visual professionals) deal with the arrangement of abstract and visual forms. There is very little in their training, day to day experience, and overall perspective that equips them or inclines them to work effectively in a medium which seeks to give meaningful visual expression to loads, equilibrium and forces.[4]

The consulting visual professionals must take the time to understand all of these issues. Otherwise, the engineer and his or her advisors will be constantly at odds with each other, and the design will suffer. For the group to be effective, all must share the engineer's goal of giving "meaningful visual expression to loads, equilibrium, and forces" with efficiency, economy, and elegance. An urban designer, architect, or artist can have a positive impact as an aesthetic advisor and critic, but the engineer must have the last word.

Organizing and controlling such a group requires that the engineer develop sufficient knowledge about aesthetics and have sufficient self-confidence to recognize valuable ideas and reject inappropriate ideas. Engineers are sometimes handicapped in this process because they are used to making decisions based on calculations and tests that seem to admit just one answer. But most engineers realize that, for any given situation, there will be a large number of concepts that will satisfy analysis, and the designer's goal must be to choose the one that best fits the situation (Figure 3-27).

Some have observed that the public seems to more readily accept bridges designed by teams that include architects, urban designers, or landscape architects. People may feel that more of their goals will be met when such professionals are involved, probably because most people in these professions are skilled at speaking to and responding to community concerns. Unfortunately, engineers have a reputation for being insensitive to community wishes, due in part to many engineers' inability to speak clearly and knowledgeably in this area. The engineer needs the vocabulary and knowledge to be effective as the project's spokesperson to the client and community groups, even concerning aesthetic ideas. Successful visual professionals are articulate spokespersons for their ideas. Lacking the appropriate presentation skills,

FIGURE 3-27 *This pedestrian bridge for I-235 in Des Moines, Iowa, was designed by engineer JMI International (now Earth Tech) with aesthetic advise from architect Miguel Rosales.*

the engineer may well see a more articulate individual in another profession take over his or her leadership role. Gaining the vocabulary and knowledge to respond to a community's aesthetic concerns allows an engineer to fulfill the leadership role and retain the community's confidence.

The final concern for involving other visual professionals is apportioning credit. Many engineering firms see a project as the product of their *firm;* they do not recognize even individual staff members in their own company, let alone their subconsultants, who may have had a significant influence on the final result. It would be fairer to emulate the tradition of architectural firms, which often recognize individual contributors within their own firm, as well as subconsultants. The opposite problem arises in the way the general press treats bridge design teams. Architects lead the design team for buildings, and the general press assumes that role applies to bridges as well. For this reason, architects working on bridges sometimes are given more credit than they deserve by the press. For bridgework the term *aesthetic advisor* more correctly describes the role, and should be substituted for *architect.*

BRIDGE AESTHETICS IN THE ENVIRONMENTAL IMPACT STATEMENT

A major bridge is always an important feature in a landscape or within a community (Figure 3-28 on page 94). Numerous people see it from many different locations at many times of the day or night. People going to and from work may see it twice a day. They experience it as a well-remembered event of their daily routine. What a major bridge looks like is always a significant issue to its community. Even the appearance of smaller bridges in important locations can become an issue.

When the time comes to replace a major bridge or build a new one, the first step is to do an environmental impact statement (EIS) or environmental assessment. This study must typically confront a series of difficult questions:

- Should a new bridge be built?
- Should an existing bridge be rebuilt?

- Should any part of an existing bridge be saved?
- Where should the new bridge be built?

Before they agree to answer these questions, the public always wants to know, "What will the new (or rebuilt) bridge look like?" Its appearance is always an issue. Unfortunately, the rules for the preparation of environmental impact statements seem to discourage consideration of bridge appearance. In order to discuss appearance, you have to be able to show examples of what a bridge would look like at that site. Examples from other places are not good enough. You cannot take a picture of the Sunshine Skyway (Figure 1-3, page 2) and say to the public, "We can build you one just like it over the Potomac River (Figure 7-7, page 295), except it will be wider and not as long or as tall." Visualizing the bridge that would result from that statement requires an extreme leap of imagination. The public cannot make that leap; most engineers can't either. A visualization must be made of a bridge as it would look at that location. That would seem to require you to do some engineering to get the bridge sized and placed correctly. And that is the problem. Bridge engineering studies are considered preliminary engineering, and the rules do not permit preliminary engineering until after the final EIS is complete. So, how can you show the public what a bridge alternative would look like if you are not allowed to draw a bridge?

One answer is that you don't. You tell the public that bridge type studies will be done later, and that you cannot show them anything at this time. But that approach can bring a project to a standstill. Because they can't discuss their concern about appearance directly, the public goes searching for surrogate issues through which they can. Noise, air pollution, or historic preservation may become controversial, while the actual underlying concern is appearance. The public also develops worst-case images in their minds, bridges that tower over their communities or obliterate neighborhoods.

The better answer is to find a way to introduce discussions of appearance into the preparation of an environmental impact statement. This can be done by illustrating several generic bridge types sized to fit the site. The examples have little engineering content beyond location sizing, so there is no concern about doing preliminary engineering too soon. Presenting several different bridge types also avoids raising the suspicion that a single alternative has been selected prematurely. To keep the process clear in everyone's mind, the examples are specifically identified as possibilities only, and it is clearly stated that the actual bridge type will be determined in a bridge type study, to be done after completion of the final environmental impact statement.

The illustrations are inserted in photographs of the site, taken from as many viewpoints as necessary to respond to public concerns. If the public is concerned about what the bridge will look like from the town park, insert an illustration of the bridge into a photograph of the bridge site taken from the town park. If the public is concerned about what the bridge will look like from the local retirement community, insert an illustration of the bridge into a photograph of the bridge site taken from the local retirement community. Helicopter shots are not enough. The background photos must be from locations that are frequented by the community every day. To keep the whole process honest, the bridges illustrated must have realistic dimensions, be accurately drawn, and be within the expected budget for the project. And the examples must themselves be attractive; otherwise, all of the potential positives can be undone by the public's reaction to an ugly bridge.

Once people can visualize what the bridge options will look like within the actual, familiar setting, their impressions become more realistic. Visualizations ensure that the final decision will be based on a realistic understanding of the visual effects of the project, not worst-case imagination. More important, everyone who sees the visualizations will have a similar impression. These impressions become part of a common language that allows people to share their opinions, confident that they are talking about the same thing. If the design team responds to the resulting comments with an open-minded attitude of active listening, people begin to believe that they can have a real role in the development of the project and that their views will be respected.

Once a project succeeds in reaching that point, several very positive effects occur. When people can visualize and understand the options, when they believe that you are going to listen to them, they gain confidence in the study. They begin to focus on appearance and discuss appearance, and surrogate issues become less of a problem, and may even go away. An even more important result is based on this fact: People understand that building a major bridge is an important creative act. Once they believe that you are going to listen to them, they realize that they will be participating in this creative act, and they become positively energized and excited by the opportunity. Their negative concerns turn to positive enthusiasm. Many change from being opponents to being supporters.

This approach is not a panacea. There are some issues and some people who will not yield to it. However, getting the public positively engaged in a project and removing surrogate issues narrows the range of remaining issues and opponents.

The community dialogue that occurs during the EIS process often results in a consensus about the qualities that the new bridge should have. This can become the basis for the design intention/vision statement, and the project can move quickly into conceptual engineering.

CASE STUDY

I-74 over the Mississippi River Joining the Quad Cities

A current study of a new bridge for I-74 crossing the Mississippi River and joining the Quad Cities of Moline and Rock Island, Illinois, with Davenport and Bettendorf, Iowa, is an example of how to

FIGURE 3-28 *The existing twin suspension bridges carrying I-74 across the Mississippi River between Davenport, Iowa, and Moline, Illinois, are considered symbols of the entire Quad Cities region.*

FIGURE 3-29 *The existing I-74 bridges seen from an upstream park.*

incorporate bridge aesthetics into an environmental study.[5] The existing crossing consists of two small, almost identical suspension bridges built side by side (Figure 3-29). The bridges are narrow, barely wide enough for two lanes, and neither has shoulders. But the community considers the bridges well-loved and distinctive landmarks for the area and is reluctant to see them simply disappear. The study involves the question of whether to replace one or both and whether to build the replacement on the existing alignment or just upstream. One alternative envisions keeping one of the existing suspension bridges as a pedestrian/bicycle facility.

To illustrate the range of possibilities, sketch concepts for an arch bridge, suspension bridge, and cable-stayed bridge were developed. The concepts were shown on background photographs of the area taken from key locations. The bridges were shown by themselves on the two alternative alignments or in combination with one of the existing bridges. Figures 3-30.1 to 3-30.4 show the three all-new options

FIGURES 3-30.1 THROUGH 3-30.4 *Visualizations comparing an arch bridge, a suspension bridge, a cable-stayed bridge, and a combination of a new suspension bridge with one of the existing bridges serving as a bicycle/pedestrian bridge, enable residents to understand what new bridge options would look like compared to the existing bridges.*

and a combined option as they would be seen from an upstream park. After viewing the illustrations, the community quickly realized that a new bridge would visually overwhelm the existing bridge, and the option that included one of the existing bridges lost support. The community also realized that the tower of a cable-stayed bridge would be much taller than all of the other options or the existing bridges, causing many to have second thoughts about that bridge type. Visualizations of the driver's views (Figure 2-29, page 44) were also done to give frequent bridge users, who include a large percentage of the Quad Cities population, an idea of how their daily experience would be affected. Taken together, the visualizations made it clear that any one of the bridge types could create a new but equally memorable symbol for the Quad Cities area.

1   Menn, C., 1998. "Generelle berechnung [general calculations]," Schweizer Inge-niuer und Architekt, Lausanne, Switzerland 44 quotation translated from the German by D.P. Billington).

2   Lockard, William Kirby, 2003. *Design Drawing,* New York: W.W. Norton & Company.

3   The author was aesthetic advisor to Burgess and Niple for the design of the High/Main Street Bridge, Hamilton, Ohio.

4   Gauvreau, Paul, 2003. *The Three Myths of Bridge Aesthetics,* Toronto: University of Toronto.

5   Rosales Gottemoeller & Associates provided visualizations, urban design, and visual analyses for the I-74 study, as part of a team led by CH2MHill, working for the Iowa Department of Transportation.

chapter four

A Design Language: Guidelines for the First Five Determinants of Appearance

"Perfection is finally achieved not when there is no longer anything left to add, but when there is no longer anything to take away. It is as if that line which the human eye will follow with effortless delight were a line that had not been invented, but simply discovered; had, in the beginning, been hidden by nature and, in the end, been discovered by the engineer."

—Antoine de Saint Exupery, French Aviator and Author

This chapter and Chapter Five "take bridges apart," to look at the appearance of each element: superstructure, piers, abutments, parapets and railings, surface colors, textures, and ornamentation. The guidelines address problems that appear over and over again in bridge design.

This is not meant to imply that bridges can be designed in pieces. Each element must be considered in relation to the whole. The guidelines should be thought of as a type of language. Just as in a written language, the guidelines can be combined in countless ways to arrive at a statement that is uniquely suited to each structure.[1] In Chapter Six we will analyze example structures by showing how the guidelines have been applied, or not applied, in more or less successful combinations.

The guidelines in this and Chapter Five are organized around each Determinant of Appearance; specifically, they are numbered according to the Determinants of Appearance for reference in later discussions. This chapter looks at those determinants that are usually thought of as purely engineering decisions, even though they have the most important impact on the appearance of the bridge. Chapter Five covers those determinants in which aesthetics are usually thought to have more of a role, even though they have less influence on appearance than the first five.

The guidelines describe solutions for each Determinant of Appearances that have worked well in existing bridges, as judged by the reaction to these bridges of engineers, visual professionals, and the public. However, they should be seen as hypotheses—subject, as in science, to continual review and potential change as technologies, social needs, and aesthetic ideas change. Each guideline is coupled, whenever possible, with a principle of perception or other factor that underlies it. This gives the engineer the opportunity to judge the guideline for him- or herself and to modify it without losing the principle upon which it is based. The guidelines are stated in a general way, so that the engineer can adapt them as appropriate to individual preferences and local conditions.

GEOMETRY

This is the first and most important Determinant of Appearance. It sets the basic shape of the bridge and its relationship to surrounding topography and other structures.

The bridge engineer is often presented with the geometry as a finished product, with little opportunity to influence it. And because geometry is the most important Determinant of Appearance, this situation can present the bridge engineer with insurmountable aesthetic problems. He or she must seek to influence the geometry in ways that will improve the chances for aesthetic success. Often, small adjustments in the horizontal and vertical alignments can make enormous improvements in appearance with no appreciable effect on safety, cost, or particular clearances or control points. The question should always be asked, and the analysis made, just to make sure no opportunity is missed.

Position and Alignment

Small adjustments in the position of the roadway relative to the features around it can make the engineer's job much easier by eliminating the need for complex pier designs or simply placing the structure in a more logical position in the landscape.

Look for the Shortest Apparent Distance between Points

This sounds too obvious to consider. However, sometimes the bridge engineer's analysis of the site shows that the highway alignment creates a crossing that does not *look* like the shortest distance across the obstruction. In these cases it is sometimes possible to make small shifts that allow the bridge, for example, to spring from a topographic promontory, or to take advantage of a protrusion of the shoreline.

Adjust the Horizontal Alignment to Simplify Pier Placement

One of the most difficult aesthetic problems is dealing with multiple pier designs within the same structure, such as mixing multicolumn frames with hammerheads. This can be particularly difficult for viaducts with changing ground plane requirements below. Often, the need for multiple pier types can be reduced or eliminated by making changes to the horizontal geometry without affecting cost or safety factors.

In Interchanges, Look for Ways to Simplify Bridge Configuration and Pier Placement

Interchange layout tends to be driven by traffic engineering and the geometrical optimization of each ramp. This often results in bridges with decks that widen and branch, and illogical span arrangements caused by restrictions on pier placement. It is difficult to make the resulting bridges attractive, so it is worth the effort to investigate possible refinements that reduce or eliminate these conditions.

FIGURE 4-1 *The original plan for the Martin Luther King/Fleur Drive interchange in Des Moines, Iowa, required five bridges, two of them branched, all on different profiles, to make two crossings of the Racoon River.*

FIGURE 4-2 *The revised interchange plan allows all of the roadways to be accommodated on two bridges, thereby creating the opportunity for distinctive bridge designs.*

The Martin Luther King Parkway/Fleur Drive Interchange in Des Moines, Iowa, is an example of how this can be done. The original plan resulted in unnecessarily complicated configurations for the two crossings of the Raccoon River (Figure 4-1). The northern crossing (the upper crossing in the photo) would have required three separate structures, each at a slightly different elevation, one of them branched. The eastern crossing would have required two separate structures at different elevations, each branched. For several reasons, including community opposition to the imposition of such a complicated interchange on its riverside park, a comprehensive restudy of the interchange was undertaken. As a result, the northern crossing was refined into a single structure of constant width carrying two three-lane roadways. The southern structure was also refined into a single structure of constant width carrying two two-lane mainline roadways and one two-lane ramp (Figure 4-2). This made possible the design of a structure using unbraced through arches occupying the spaces between the roadways. The arches frame the downtown skyline for travelers approaching from the airport, and create a gateway for Des Moines[2] (Figure 4-3).

FIGURE 4-3 *The eastern crossing of the Racoon River is a double unbraced arch, which will create a gateway on the route from the airport to downtown.*

The Internal Harmony of the Alignment[3]

Because of the size of modern roadways, the pavement itself occupies a large part of the visual field. It's form can be attractive or unattractive, depending on its geometry. For long bridges, particularly those over water, the form of the pavement can be seen as a "ribbon in space" by viewers of the bridge. The attractiveness of its form will strongly influence the aesthetic impression the bridge makes (Figure 4-4). The following guidelines will help assure that the form is attractive.

Construct Horizontal and Vertical Alignments from Long, Continuous Curves

Vertical and horizontal alignments made up of long, continuous curves and short tangents will look better than alignments made up of short, discrete segments. Because of the extreme perspective foreshortening that takes place in the highway environment, short curves will look like kinks in the alignment. Because we expect to find gradual changes in direction in the highway environment, especially when traveling at high speed, an alignment that appears kinked and thus seems to require a sudden change of direction will look disturbing and inappropriate.

Closely spaced curves in the same direction should be connected with compound transition curves, without an intervening tangent. Reverse curves should have only enough tangent to provide for a graceful superelevation reversal. The best way to reverse curves is with spiral transition curves or compound curves that approximate spirals.

Coordinate the Locations of Horizontal and Vertical Curves

Engineers judge horizontal geometry based on plan drawings, and vertical curves based on profile drawings, and forget that the viewer sees neither. In the three-dimensional reality seen by the viewer, the two appear together and influence the appearance of each other. If the horizontal and vertical curves do not overlap at all, there is usually not a problem. If they overlap entirely, forming a three-dimensional curve in space, they are usually quite attractive. Problems arise when a horizontal curve half-overlaps a vertical curve. The result looks like a kink in the roadway. The effect is worsened for long bridges over water or long, high ramps, because a long length of the structure is visible all at once. Designers should start by plotting the location of the horizontal curves on the profile (Figure 4-5), then try to

FIGURE 4-4 *The too-short curve on the left ramp looks like a kink from many viewpoints.*

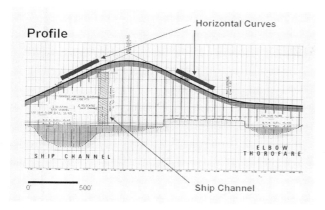

FIGURE 4-5 *By plotting the location of horizontal curves on the original profile for New Jersey's Route 52 Causeway Replacement, it was easy to see that the horizontal curves and vertical curves were not well coordinated.*

FIGURE 4-6 *The upper visualization shows the kinked appearance that would have resulted from the original geometry shown in Figure 4-5. The revised geometry shown in the lower visualization eliminates the kink by placing the horizontal curves in the same position as the vertical curves but with a somewhat longer length.*

adjust the location of both horizontal and vertical curves to improve the coordination. Highway engineers will often claim that this is impossible, but under scrutiny it usually turns out that enough flexibility is available to make the refinement. Figure 4-6 shows the more attractive flowing alignment that can be achieved.[4] Views like this, and even drivers' view animations, are easy to create with the 3D modules of common CAD programs like MicroStation, and they are excellent checks on the quality of the geometry.

Relate Curve Length to Structure Length

How long a curve needs to be is a question that can only be answered in the context of the specific bridge. To start with, a curve that looks good is almost always substantially longer than the minimums set by the American Association of State Highway Transportation Officials (AASHTO). Second, the curve length should bear some relation to the length of the bridge. For bridges less than 2,500 feet long, with one horizontal or vertical curve, the curve should be no less than half the length of the bridge.

Analyze the Effect Superelevation Transitions Have on Parapet Alignment

An analysis needs to be made of the effect of superelevation transitions on the parapets. Because the top of the parapet is near driver's eye height, and because of the foreshortening effect of perspective, the horizontal and vertical alignment of the parapet is very visible. In bridges with shoulders and/or sidewalks, the effect of superelevation transitions can be exaggerated. Vertical curves can add more complication. On any bridge with multiple curves, the parapet profiles should be plotted at an exaggerated vertical scale and checked for kinks and awkward curvatures.

Where Possible, Use a Crest Vertical Curve on Overpasses

Even a very slight crested vertical curve will give a bridge a slightly arched appearance, which most people find appealing (Figure 4-7).

FIGURE 4-7 *The crest vertical curve and bright color combine to create a memorable experience on the Pennsylvania Turnpike (see color insert).*

FIGURE 4-7 *The crest vertical curve and bright color combine to create a memorable experience on the Pennsylvania Turnpike (see color insert).*

SUPERSTRUCTURE TYPE

Superstructure type is usually heavily influenced by the conditions of the site and the required spans. Given the proportion of the bridge cost associated with the superstructure, its type will usually be determined by combined economic and structural considerations. However, there are situations in which two or more different superstructure types are comparable on economics and performance, and appearance can come into play. Or, situations can arise in which a different superstructure type will provide a major improvement in aesthetic quality at a small increase in cost. Finally, there are many variations of layout and arrangement within a given structural type for which improvements in appearance can be achieved at little or no increase in cost.

This book concentrates on the 60- to 500-foot span range, where girders are the most usual type of superstructure. However, rigid-frame bridges and arches are also possibilities in this span range for the right site (Figure 4-8). Trusses and cable-supported bridges have also been used, though more rarely and generally under special conditions.

FIGURE 4-8 *A through arch at Chicago's Dammen Avenue bridge with its thin floor system minimized the regarding of the street approaches (see color insert).*

Most superstructure types are built in both steel and concrete, so the choice of material by itself does not determine structural type. However, considerations of material choice and structure type are intertwined. For example, certain types, such as box girders, are generally easier to build in concrete than in steel. Material choice is discussed in more detail later. Finally, this section concentrates on the aesthetic aspects of the decision. It assumes a prior understanding by the reader of the structural possibilities and the economics of the various options.

Influences on the Choice of Superstructure Type

There are many factors that influence the choice of superstructure type. All must be understood and weighed together in order to arrive at the best solution.

- The geometry of the roadway to be carried exerts a major influence, particularly if it tapers or is curved in plan. Structural types that can be curved or tapered to match offer an overriding simplicity (Figure 4-9).

- The span requirements and required vertical clearances will determine the proportions and size of the bridge, as well as set the boundaries on what is economical and even what is physically possible.

- The nature of the topography and foundation conditions at the site will be major influences. For example, arches and rigid frames look best in steep ravines, valleys, and deep cuts, where the arch/frame can spring from the hillsides, or crossing rivers with steep banks or river walls. They also work better structurally because of the capability of such locations to resist the horizontal forces imposed by an arch/frame.

- The location of important viewpoints and nearby land uses should influence the structural type. For example, the underside of a structure over an urban street or park will be visible to many people. I-girder bridges are difficult to make attractive from the underside because of the visually complicated bracing

FIGURE 4-9 *This bridge, with chorded precast concrete I-girders, demonstrates the variable slab overhangs and the visual confusion that can result from trying to make a straight structural element fit a curved structure.*

details, and because the fascia girders shadow all the other girders. The underside becomes a dark, complicated, and somewhat threatening space inhabited by pigeons and worse.

- If the structure is intended to carry a symbolic role, that can most effectively be satisfied through the choice of structural type. For example, if it is desirable for the structure to frame an important view, then a structural type with a curved, arched soffit will be an effective choice (see Figure 1-8, page 4).

Effects of Material Characteristics

Usually, material choice is heavily influenced by the cost of the material in the locality, the availability of experienced contractors and fabricators, ease of construction at the specific site, and environmental factors such as the presence of a corrosive environment. However, the choice can also dramatically affect the appearance of the structure, and these effects should be considered.

The choice of superstructure material can affect appearance in four ways: apparent thinness, shape, details, and color. The typical precast girder is significantly deeper than a steel welded-plate girder of equivalent span. This difference in depth has a major effect on the superstructure's overall appearance. However, by the use of appropriate post-tensioning, poured-in-place, and precast segmental concrete, structures can be made almost as thin as steel structures.

Cast-in-place concrete girders can be built with a smooth surface of any shape, as long as the contractor can economically build the formwork. Steel girders are sharp-edged and are limited in shape by the need to fabricate them from plates and rolled sections. Standard precast girders have characteristic shapes and are limited in the ways they can be economically connected.

Differences in details enforced by the material choice need to be considered. On steel bridges, bolts, stiffeners, bracing plates, and splice plates will be visible, and their appearance needs to be considered. With concrete box girders, all details and connections are typically internal and unseen.

Concrete and A588 weathering steel have strong characteristic colors. Superstructures may need to be coated to offset the effect of material color. For example, to achieve visual continuity in a long structure, it may be desirable to coat steel box girders with a light, warm gray so they look like adjacent concrete box girders. Concrete or light-colored paint on steel is advantageous for bridges where pedestrians use the under bridge area because more light is reflected into the underbridge area.

If coating is not desirable, then the natural color of the material should be a consideration at the time of material selection. This can be particularly important when there will be significant pedestrian use of the underside of the bridge. For example, A588 weathering steel attains a dark brown–black color, which provides an unpleasant "ceiling" for a pedestrian space, hence needs to be offset with additional lighting or other measures. The situation is exacerbated by repetitive I-girders. The dark matte surfaces soak up whatever light may be available between the girders. However, if few pedestrians are likely to be use the space, then color, being lower on the scale of importance for Determinants of Appearance, is a less significant consideration for selecting the superstructure material. (See Chapter Five for more information on color.)

No one of these characteristics is in itself good or bad. All need to be considered for their support of the design intention, which may indicate a material based upon the desired structural conditions or visual concerns. For example, steel was required for the river crossing shown in Figure 1-11.2 (page 7) in order to meet design concept requirements for maximum vertical clearance while still connecting to existing streets. The choice of steel also allowed the assembled girder to be rolled into place to replace an existing truss in a minimal amount of time. Once the choice is made, the chosen material can then be detailed to meet the requirements of the design intention.

Considerations for all Superstructure Types

Certain criteria should influence all bridges. Chapter One explained that relative slenderness is an advantage, which depends heavily on the choice of structural type.

Seek Relative Slenderness When Picking Superstructure Type

Choosing continuous girders over simple spans will immediately improve the looks of the structure because structural continuity allows a thinner structure. Precast concrete I-beams, which are usually only partly continuous structurally, will look (and be) heavy compared to continuous steel girders because of their greater depth for a given span. Post-tensioned concrete slab bridges can look very light because of their minimal depth (Figures 2-24, page 42; and 4-28, page 117).

Maintain Continuity of Structural Form, Material, and/or Depth

The bridge should be a single unified concept. Changes in structural type or depth to accommodate differing span conditions should be made smoothly, and similar structural shapes, materials, and/or colors should be used to tie the structure together (Figure 4-10).

If the main span or spans require a change to a different type, such as an arch, truss, or rigid frame, then maintaining an appearance of continuity becomes more difficult. Arches and frames often are combined with a series of girder–approach spans (Figure 4-11). Combining these two different forms in one structure can seem visually confusing. A good approach is to select features of the main structure to extend over the approach spans.

FIGURE 4-10 *The tapered side-spans of this bridge smoothly continue the lines of the main span girders.*

FIGURE 4-11 *The constant-depth girder of Oregon's Alsea Bay Bridge is an element of continuity across different structural types (see color insert).*

For example, a deck stringer for an arch can be matched with a girder of the same depth on the approach spans. Deck overhangs can be kept constant from a girder span to a deck truss. The end span of the truss can be tapered to the same depth as the girder. The same color or material can be used on both types. The more consistent the elements are, the more continuous the structure will seem. If these techniques are insufficient to visually establish continuity, then a major pier or other vertical feature can be inserted between the two types of structure to visually clarify where one type of structure starts and the other stops.

Girder Bridges

For bridges with spans less than 500 feet, under most conditions and with today's technology, the most cost-effective structure is a girder bridge. The girder category includes I-girders, box girders, and voided slabs. Most types can be built in steel or concrete, and the concrete can be cast in place or precast. Table 4.1 below compares the capability of the most frequently used options to meet aesthetic goals.

Use Curved Girders for Curved Roadways

Curved girders are better than straight girders on chords for all but the gentlest curvatures. Curved girders eliminate the variable deck overhangs (and scalloped shadows) of spans on chords. They also eliminate all of the structural tinkering necessary to fit straight girders to circular curves and still meet clearance requirements and fixed pier locations (as in Figure 4-9, page 103). In more basic terms, curved girders are better because the structure itself reflects the lines of motion, which are the dominating features of the transportation environment, and which are, in fact, the structure's reason for existence.

If straight girders must be used, they should all be of a similar length so that the variations of deck overhang occur in a constant rhythm along the structure.

TABLE 4.1 *Aesthetic Capabilities of Girder Types*

	A	B	C	D	E	F	G
1	Girder Type	Material/ Construction	Curved?	Made Continuous?	Uses Integral Pier Caps?	Attractive Underside?	Widened or Branched?
2							
3	I-Girders	Steel	Yes	Yes	Yes	No	Difficult
4		Precast Concrete	No	Difficult	Difficult	No	Difficult
5	Box Girders	Steel	Yes	Yes	Yes	Yes	Difficult
6		Cast-in-place Concrete	Yes	Yes	Yes	Yes	Yes
7		Precast Concrete	Yes	Yes	Yes	Yes	Difficult
8	Voided Slabs	Cast-in-place Concrete	Yes	Yes	Yes	Yes	Yes

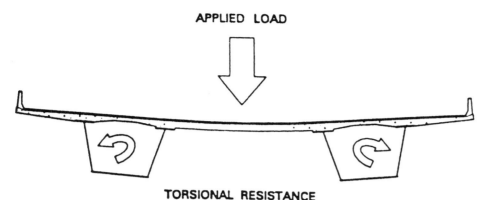

APPLIED LOAD

TORSIONAL RESISTANCE

FIGURE 4-12 *The torsional stiffness of box girders improves load distribution and allows wider spacing than for I-shaped girders.*

Consider Box Girders or Post-Tensioned Concrete-Voided Slabs for Prominent Bridges, Bridges with Curvature, Bridges That Widen or Branch, or Bridges Where the View of the Underside Is Important

Box girders, both steel and concrete, have been used in prominent locations because of their relative thinness; the clean, simple, attractive appearance of their undersides (the stiffening and bracing details are mostly hidden); because their flat underside reflects light into the underbridge area; and their torsional stiffness often allows for thinner piers and more flexible pier locations (Figure 4-12).

The design of steel box-girder bridges has been hampered by a tendency to use too many boxes. The bridges have been treated like plate-girder bridges with adjacent bottom flanges connected. This approach misses the opportunity to take full advantage of the torsional stiffness of box girders and adds unnecessary cost and visual complication. Designers should seek to use as few widely spaced box girders as reasonable.[5]

A recent study in Nebraska using high-performance steel has shown that two steel box girders are sufficient for a two-lane roadway and can be more economical than precast I-girders, even for relatively short spans.[6]

Precast concrete segmental box girders have generally been used only for relatively long bridges because the cost of building the specialized forms must be amortized over a large project. However, a range of standard bridge segmental designs has recently been developed by the American Association of State Highway and Transportation Officials.[7] Once the standardized forms are made available in precasting yards, this type of construction will become economical for smaller projects. Precast concrete segmental box girders are especially applicable to the long flyover ramps of major interchanges, where they have structural, constructability, cost, and aesthetic advantages (Figure 4-13). (See also the case study of the Big I Interchange in Chapter Six.)

Consider Abutment-Restrained Girders When Vertical Clearance Is Minimal or When It Is Desirable to Frame a View through the Structure

Three-span girder bridges can be built with the end spans hidden within the abutments. The center span can be treated as a fixed-end girder, with maximum moments at the abutment. The resulting structure allows longer spans than a simple supported girder, and allows the mid-span depth to be reduced. Figure 1-8 (page 4) shows how abutment-restrained girders can be thin at mid-span, their curved bottom edges effectively framing the views beyond.

FIGURE 4-13 *The Fort Lauderdale–Hollywood International Airport Interchange is one of the first uses of the AASHTO standard precast concrete box girder segments. It won its designers, Beiswenger, Hoch & Associates, the 2004 Engineering Excellence Grand Award of the Florida Institute of Civil Engineers.*

FIGURE 4-14 *The curved lines of this bridge with integral pier caps flow past the piers and are continuous from one end of the bridge to the other.*

Consider Integrally Framed Cross Girders

Integrally framed cross girders have also been used in high-exposure locations because they minimize the size of the pier, provide for more flexible pier location, and emphasize the visual continuity of the superstructure (Figure 4-14).

Integrally framed cross girders can be particularly effective in improving the appearance of I-girder bridges, both steel and precast concrete. If carefully detailed, they can provide structural redundancy and need not add significant cost.

When Adding Girders to Accommodate Splits and Widenings, Do So in a Clear, Systematic Manner

Splits and widenings to accommodate ramps can result in a hodgepodge of half-girders and odd spacing, particularly when working with steel or precast concrete I-girders (Figure 4-15).

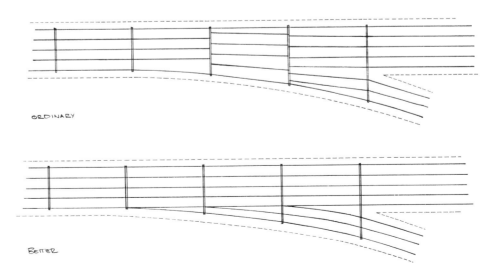

FIGURE 4-15 *The method of adding girders at a ramp junction should present an easily understood pattern.*

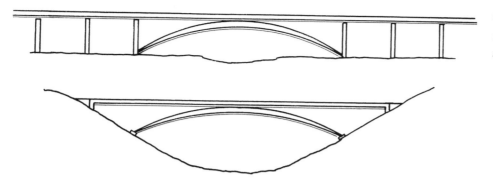

FIGURE 4-16 *An arch bridge needs a site that contains the structural and visual thrust.*

Arches/Frames

Arch and rigid-frame bridges less than 300 feet in span are rarely economical with today's technology. However, there are sites where restrictions in floor system depth have made through arches cost-effective. In any case, arches and frames continue to have strong visual appeal because of their shape. They are definite possibilities for the right site.

Arch and Rigid-Frame Bridges Look Best Where the Surroundings "Contain" the Visual Thrust of the Arch

Even tied arches look best contained in river valleys with high bluffs (Figure 4-16). Thrust blocks can provide the same sense of "containment" if they are large enough (Figure 4-17).

Rigid-Frame Bridges Look Best with Legs That Are One-Quarter to One-Half the Span Length

Rigid-frame bridges need height to allow for a graceful length of leg. Rigid frames with short legs tend to look stubby (Figure 4-18.1 and 4-18.2). These considerations are less of a problem where the legs are enclosed triangles (Figure 4-19), when the bridge can fit into most sites. Rigid frames can be very slender at mid-span.

FIGURE 4-17 *An arch was selected for Vancouver, British Columbia's Big Qualicum Bridge because of the site. The design provides plastic hinges in the concrete arch rib to control earthquake stresses.*

FIGURE 4-18.1 AND 4.18.2

The importance of height to a rigid-frame bridge.

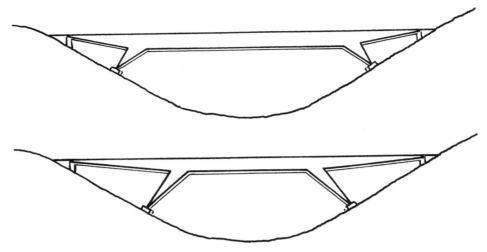

They will be visually more successful the more the elements are shaped to reflect the change in stresses across the structure. (See also Figures 2-3, page 29; 2-25, page 43; and Figure 6-10.3, page 204).

Trusses

In recent decades, truss bridges have rarely been used at this span range because of cost, maintenance, and appearance. However, there is now renewed interest in truss bridges for local roads. Also, prefabricated truss bridges are often used as pedestrian bridges. Finally, several concepts for composite and pre-stressed trusses are in the experimental stage, and the availability of high-performance steel may lead to more economical trusses. It is likely that there will be more serious interest in trusses for shorter-span bridges in the years ahead.

The keys to good appearance in truss design are:

- A graceful overall shape

- Simplicity: minimum number of members

- Consistency: All the members should be at consistent angles. For example, a truss with all diagonals at a 60-degree angle will look better than a truss that mixes diagonals and vertical members. The diagonals will look like a continuous zig-zag line.

- Small and attractive connection details

FIGURE 4-19 *This triangular frame looks good even in a site with a low clearance.*

Cable-Supported Structures

Pedestrian bridges offer some possibility for suspension and cable-stayed designs in this span range. These possibilities are discussed in Chapter Six in the "Pedestrian Bridges" section.

Vehicular-suspension and stayed-girder bridges are generally economical only for spans in excess of 500 feet. Though a discussion of these bridge types is beyond the scope of this book, it is pertinent to point out here that, as Menn showed in his Sunniberg Bridge (Chapter Two), and as shown in the bridge in Figure 3-3 (page 63), shorter-span cable-stayed bridges sometimes are the best answer to specific situations.

PIER PLACEMENT

Pier placement must respond first to under-roadway clearance requirements, hydraulic requirements, navigational channels, and foundation conditions. Visual criteria should be added to these considerations. Visual criteria fall into two categories: providing a pattern of piers that is visually logical both within the structure and relative to nearby topographic features, and providing for desirable lines of sight through the structure.

General Pier Placement Guidelines

With Few Exceptions, There Should Be an Odd Number of Spans
With an even number of spans, a duality is produced that most people find uncomfortable (Figure 4-20). Usually, there is something to be crossed: a channel, river, roadway, or ravine, and people expect that "something" to occupy, or appear

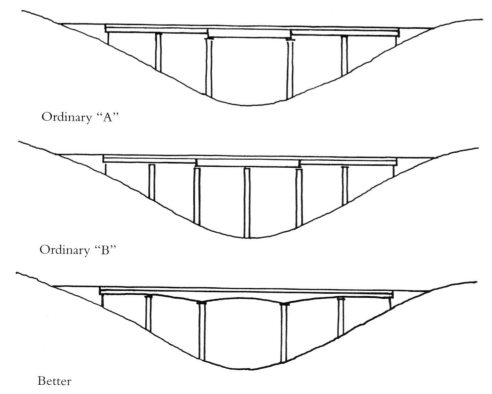

Ordinary "A"

Ordinary "B"

Better

FIGURE 4-20 *Schematic A suffers from a discontinuous superstructure. Schematic B has several problems: an even number of spans and a pier at the deepest point of the valley, as well as a discontinuous superstructure. Compare with the schematic at the bottom.*

to occupy, the center of the structure. A pier in mid-river or mid-roadway is an obvious conflict.

However, there are important exceptions to this guideline. Wherever there is a strong symmetry in the area being spanned, with an obvious pier location at its center, then an even number of spans is preferable. Frequent examples are a typical freeway with dual roadways, or a river with an island in the middle. Also, for very long bridges with more than seven or eight spans, the fact that there is an odd or even number of spans is not noticeable.

Place Piers in Logical Relation to Topographic Features

Placement with regard to topography and shorelines can be critical.

- Do not place piers at the deepest part of a valley or man-made cut. This position will produce the tallest possible pier. In addition to the cost problem, observers will find this obvious lack of logic disturbing.

- Place piers on natural promontories (Figue 4-21). This approach will produce shorter piers and will look more logical than a pier placed next to a promontory.

- Place a pier on land at or near the shoreline, more or less symmetrically placed with regard to both shorelines. Common sense calls for a pier (or abutment) on land at or near the shoreline. Deviations from this guideline should have a visually obvious reason to explain them, such as a shoreline roadway.

For Arch and Rigid-Frame Bridges, Springing Point Placement Is Critical

Place the arch or rigid-frame springing points on topographic features that appear to resist the horizontal, as well as the vertical, reactions.

- For arches and frames over water, place the shore-side springing point on shore near the shoreline to give the structure a strong visual end point to the diagonal line of the arch or frame. If the arch ends in the water, a large foundation element should be created to provide a strong visual end for the arch and visually contain its horizontal thrust.

FIGURE 4-21 *This rigid frame starts at a logical point, the benches of this man-made cut, which visually contain both the horizontal and vertical reactions.*

- For both arches and rigid frames in hilly topography, try to find natural locations on the valley walls (such as a rock outcrop) or cut sidewalls (as on a bench) to place the springing (Figure 4-21).

Consider Superstructure Costs and Foundation Costs When Developing Optimum Span Lengths

Typically, superstructure costs increase and foundation/pier costs decrease as spans get longer. These relationships can be plotted as curves showing the change in element cost at various spans. By adding the two curves together, the true effect of various spans lengths can be determined. Figure 4-22 shows an example of a combined cost curve.

Often the combined curve does not show a clear minimum. The only logical conclusion is that the cost effect of longer spans is negligible over a significant range. In such cases, span lengths should generally be set at the long end of the range. Fewer piers will mean a more open view through the structure, particularly at oblique angles.

In General, Span Length Should Exceed Pier Height

Throughout the bridge, it is the *relative* sizes of the major elements that have the strongest effect on the visual impact. For pier placement, the key proportion is span versus vertical clearance, or a better way to look at it, span versus the overall shape of the space beneath the bridge. Generally, the bridge will look better the more the horizontal dimension of this space (the span) exceeds the vertical dimension (Figures 4-23.1 and 4-23.2).

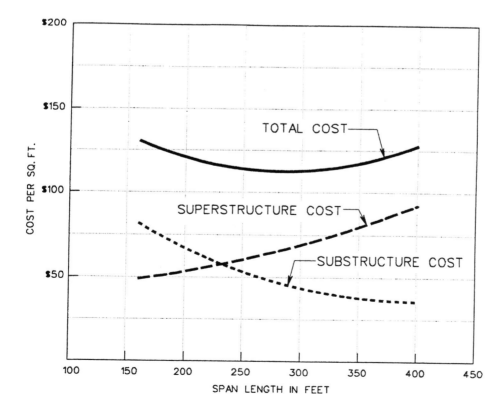

FIGURE 4-22 *A typical cost curve; the curve changes with each project.*

FIGURE 4-23.1 AND 4-23.2

Emphasizing horizontal proportions in pier placement.

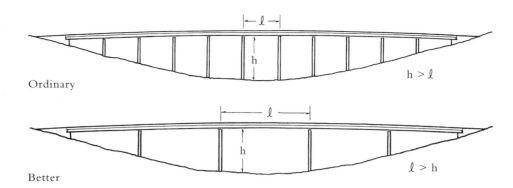

Most bridges are in an environment where the horizontal dimensions of the bridge itself and its surroundings are much greater than the height of the bridge. The longer the spans, relative to pier height, the more the bridge will seem to be consistent with its environment. Also, logic says that the reason for the bridge is to cross something, and additional piers only get in the way. This idea can reach a point of diminishing returns when the pier and girder elements become so massive as to overwhelm the site. With very high bridges, economics may make it unfeasible to make the span significantly greater than the height (Figures 4-24.1 and 4-24.2).

Ratio of Height to Span Length Should Be Similar from Span to Span

Once the ratio of span to height (l/h) has been established for the main span, the secondary spans should vary in proportion to their individual heights in order to maintain the l/h ratio approximately constant for all spans. The bridge will have greater unity if these *proportions* are consistent, regardless of span lengths.

Open Up Spatial Corridors through the Structure

The width and continuity of spatial corridors through the bridge should be as generous as feasible (Figure 4-25). The intrusiveness of the bridge in the town or landscape will depend on the degree to which this is accomplished (Figure 4-26). It is particularly important in interchanges, where the ability to see beyond the bridge is a matter of safety as well as appearance.

As discussed in Chapter Two, actual sight lines need to be recognized. For example, a multicolumn bent will look simple when seen end-on (as in an elevation draw-

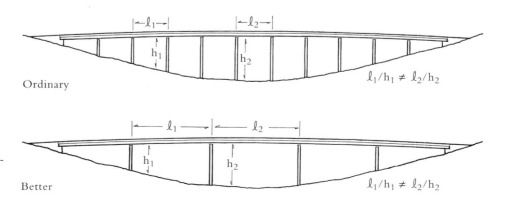

FIGURE 4-24.1 AND 4-24.2

Multispan bridges will look better with consistent proportions of span to height in all spans.

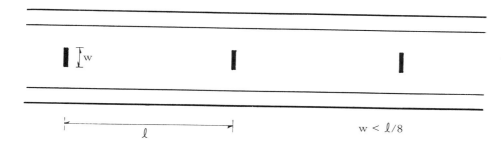

FIGURE 4-25 *Ratio of pier width at the base-to-span length for single-shaft piers.*[8]

$w < \ell/8$

ing), but the hidden columns will be very apparent in the more normal diagonal views. A group of such bents can become a "forest" when seen from the usual angles. Figure 6–33 (page 228) shows how a forest of columns can block sight lines through the bridge. Figure 6–38 (page 230) is an example of how the problem can be reduced. General rules for maintaining view corridors through structures are:

- For single-shaft piers, span lengths and overhangs should be adjusted so the pier width does not exceed one-eighth of the span length.

- If multiple column piers are planned, the span should be set so that the total width of the column group does not exceed one-half of the span length. (See also the "Pier Shape" section in Chapter Five.)

Piers for Skewed Structures

The needs of highway and transit geometry often enforce crossings of other facilities at sharp angles, or *skews*. These situations pose visual as well as structural challenges. At skew angles down to about 60 degrees, a standard girder bridge can be adapted without major effect on the appearance. However, at lower angles, piers and abutment walls become substantially elongated. These elements become large portions of the visual scene, to the detriment of the structure's overall appearance. Also, the nearer part of the superstructure appears larger to the viewer than the more distant part, and the distortion can be disturbing. These effects tend to be underestimated by engineers viewing the structure in elevation drawing, resulting in many unsightly skewed bridges. Preparing a three-dimensional driver's eye view will pinpoint the problems and suggest some solutions (Figures 4-27.1 and 4-27.2).

Addressing the problems of skew structures requires considering superstructure type and shape, pier placement, and abutment placement all at the same time, as a decision about one will have major impacts on all the others.

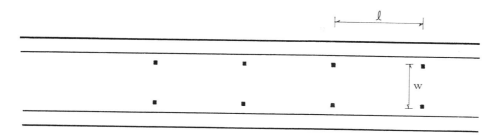

$w < \ell/2$

FIGURE 4-26 *Ratio of column spacing and span length for two or more columns.*[8]

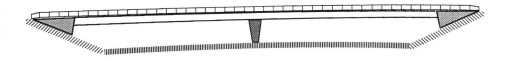

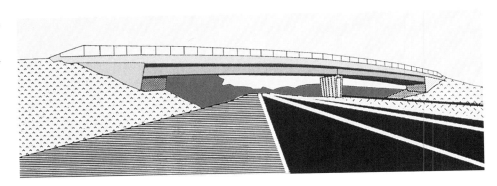

FIGURE 4-27.1 *Compare this elevation drawing of Maryland Route 18 over U.S. 50 to what the observer actually sees in Figure 4-27.2.*

FIGURE 4-27.2 *Actual appearance of Maryland Route 18 over U.S. 50; this is an abstraction of the photo in Figure 1-23.1 (page 22).*

Select Superstructure Types That Allow Large Overhangs and Few Supports

Adopting a superstructure type that has larger overhangs narrows the piers and helps compensate for the skew elongation. The desirable goal is a superstructure that can be supported on a single round or elliptical pier. Not only does this minimize the visual presence of the pier, but it also presents much more flexibility in the placement of the pier in complex interchange situations.

Box girders and post-tensioned concrete-voided slabs lend themselves to this situation because their torsional stiffness allows for larger overhangs and creates the possibility of handling the torsional reactions at the abutments only. Ontario, Canada, has built many interchange structures that take advantage of this property (Figures 2-24, page 42; and 4-28). Structures with integral pier caps with soffits that are flush with the soffits of the longitudinal girders should also be strongly considered. The piers can be much shorter than a conventional pier because the pier cap can extend over the undercrossing roadway without causing a clearance problem.

For Very Wide Structures, Use Multiple Girders with a Single Column at Each Beam

Following this guideline eliminates the visual distraction of an elongated pier cap (Figure 4-29). There is a danger in the extreme case, however: that the appearance will become confining and tunnel-like to the driver.

For Sharp Skews, Place a Girder at Each Edge of the Overpassing Roadway, Then Place Columns at the Edges of the Underpassing Roadway

This approach can result in a very open appearance. Recent concerns about the redundancy of two-girder systems have discouraged this type of structure. However, using box girders, which are internally redundant, or paired I-girders, can address this concern. (For more on skewed bridges, see the discussions of skewed bridge abutments in the following section, of superstructure shape later in this chapter, and of pier shape in Chapter Five.)

FIGURE 1-1 *The Brooklyn Bridge, a symbol of New York City and the East Coast.*

FIGURE 1-2 *The Golden Gate Bridge, a symbol of San Francisco and the West Coast.*

FIGURE 1-3 *Sunshine Skyway, a symbol of Tampa, Florida, and of modern technology.*

Figure 1-8 *A message of pride and skill, the Gunnison Road Bridge over I-70 west of Denver frames the westward traveler's first view of the Rocky Mountains.*

FIGURE 1-23.1 *Maryland 18 over U.S. 50, one approach to a two-span bridge.*

FIGURE 2-4 *Bright color clearly differentiates the elegant load-carrying tubular arch from the deck it supports. Jorg Schlaich's Pragsattel I pedestrian bridge, Stuttgart, Germany.*

FIGURE 2-15 *One way to achieve unity is to design all elements with the same family of shapes, with the same standard details, and the same colors. The Big I Interchange, Albuquerque, New Mexico.*

FIGURE 2-41 *Christian Menn's Sunniberg Bridge near Klosters, Switzerland.*

FIGURE 2-44 *The flared pier tops allow the stays to clear the roadway on a curve.*

FIGURE 2-45.1 *Refined elegance in a structural shape based on structural role and size.*

FIGURE 2-45.2 *Refined elegance in a sculptural shape by Constantin Brancusi.*

FIGURE 2-53 *The bright yellow color of the steel box girders makes this interchange a memorable milepost for travelers using I-440 in Nashville. This outstanding design for a complex interchange is made possible by the torsional stiffness of box girders.*

FIGURE 4-7 *The crest vertical curve and bright color combine to create a memorable experience on the Pennsylvania Turnpike.*

FIGURE 4-8 *A through arch at Chicago's Dammen Avenue bridge with its thin floor system minimized the regrading of the street approaches.*

FIGURE 4-11 *The constant-depth girder of Oregon's Alsea Bay Bridge is an element of continuity across different structural types.*

FIGURE 5-37.1 *This image and the following one show the same bridge painted in two different colors, and illustrates how the color of a structure and its background affect our perception. In this image, the orange bridge contrasts strongly with the wintertime grays and browns of its background.*

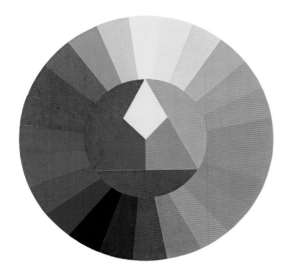

FIGURE 5-40 *The color wheel.*

FIGURE 5-41 *The center panels show tertiary colors in which no hue can be discerned.*

FIGURE 5-42.1 *Ridge Road over I-70 near Frederick, Maryland, photographed in the late afternoon in October, looking from the west.*

FIGURE 5-42.2 *Ridge Road over I-70 near Frederick, Maryland, photographed in the late afternoon in October looking from the east.*

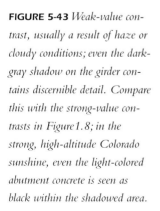

FIGURE 5-43 *Weak-value contrast, usually a result of haze or cloudy conditions; even the dark-gray shadow on the girder contains discernible detail. Compare this with the strong-value contrasts in Figure 1.8; in the strong, high-altitude Colorado sunshine, even the light-colored abutment concrete is seen as black within the shadowed area.*

FIGURE 5-44 *Photographed halfway through a repainting, the Maury River Bridge in Virginia demonstrates the interaction of color and background: The aluminum-colored span takes on the sky color and seems part of the sky, while the brown span emulates the colors of the topography and clearly ties together the two hillsides.*

FIGURE 5-45 *This red railing adds interest to, and unifies, the mottled concrete of this structure.*

FIGURE 5-46 *Using a colored inset to add a strip of bright color to a concrete structure.*

FIGURE 5-66 *Broadway Bridge, Miami, Florida. Floodlights reinforce the shape of the structural elements.*

FIGURE 6-23 *This arch creates a landmark along I-65 that signifies one of the main entrances to Columbus, Indiana.*

FIGURE 6.27.2 *The long spans and thin piers for the box girder supporting the Lower Screwtail Bridge in Sunflower, Arizona, allow the desert landscape to flow through the structure.*

FIGURE 6-30.2 *The haunched concrete box girder flows smoothly into the piers.*

FIGURE 6-31 *With very simple elements, this Tennessee structure conveys strength, continuity, and horizontality. The last river span is tapered to make a smooth transition between the depth of the main spans and the depth of the approach spans.*

FIGURE 6-44 *Clearwater Memorial Causeway, Clearwater, Florida. Long spans open up the view to Clearwater Beach and the sunset.*

FIGURE 6-45 *Blue floodlights will make the girder soffits a frame for the nighttime view, an effect multiplied by shimmering reflections in the water.*

FIGURE 6-57 *The structures for the Big-I interchange in Albuquerque permit open vistas through the interchange, with plenty of space for undercrossing roadways.*

FIGURE 6-58 *The rosy tan color and the blue parapet stripe tie together every element of the Big-I interchange.*

FIGURE 6-59 *Because of its color, the Big-I interchange seems to fit right into the Sandia Mountains and nearby development, as if it had always been there.*

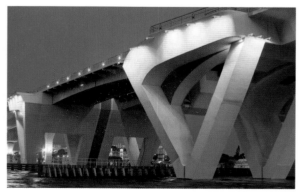

FIGURE 6-61 *Movable bridges are all about movement—marine, automobile, and the bridge itself—as this time exposure of the Duluth Aerial Bridge makes clear.*

FIGURE 6-71 *Lighting brings out the shape of the Carina Pier at night.*

FIGURE 6-76 *The apparent depth of this railroad bridge is reduced by the band of contrasting color above the horizontal stiffener.*

FIGURE 6-77 *The bridge at the Lehrter Railroad Station in Berlin uses a steel tube arch to cross the Humboldthaven barge basin.*

FIGURE 6-86 *The piers of the Rogue River Bridge in Grants Pass, Oregon, are placed to allow park uses to continue below. However, the railing color does not blend with the background, an example of the difficulty of finding a green that blends well with background foliage. A contrasting color might have been better.*

FIGURE 6-90 *The Reedy River Bridge will be lighted to be a landmark for Greenville at night as well as during the day.*

FIGURE 7-6 *The Naval Academy Bridge over the Severn River, Annapolis, Maryland. The result of a competition that resolved a major local controversy.*

FIGURE 7-11 *Entrant B's design for the Woodrow Wilson Bridge, from the Alexandria Waterfront.*

FIGURE 4-28 *This Ontario structure's post-tensioned voided slab has great tensional stiffness; torsional forces are carried to the abutments, allowing single intermediate columns, which open up clear views through the structure.*

FIGURE 4-29 *Using a single freestanding column for each girder can avoid an unreasonably long pier cap for skewed structures.*

ABUTMENT AND WALL PLACEMENT

The placement of abutments and walls in the visual field will influence the size and appearance of both superstructure and piers, as well as create elements that in themselves can be positive or negative features. The positive or negative nature of their contribution will depend significantly on where they are placed relative to other features and the topography.

Guidelines for Abutment Placement

The abutment's function is to get the bridge started at one end and bring it back down to the ground at the other end. Its visual job, as well as its structural job, is to mediate between earth and structure. The "right" decision for an abutment depends on the designer's design intention for the structure.

One-span bridges are the simplest structures. Abutment placement is *the* key element in determining the overall proportions of the bridge. For two-, three-, and four-span structures, the abutments are a lesser part of the total structure, but are still major elements because both abutments are seen at once, and frame the structure. For these bridges, and particularly for highway overcrossings, the specific placement of the abutment is crucial to the appearance of the bridge.

The range of possibilities is demonstrated by the history of highway overcrossing design. The design of highway overcrossings starts with the required clearance envelope of the underroadway. In the early days of highway bridge building, that was the limit of the bridge, and bridges looked like the drawing in Figure 4-30.

Designers soon realized that high walls create an uncomfortable degree of enclosure for motorists, cut off the view through the structure, and pose safety hazards. So they added piers at the shoulder edges and moved the abutment to the top of the slope, as shown in Figure 4-31.

While this was a big improvement, the added void areas on either side are relatively small and cut off from the major space by the piers, while the piers themselves

FIGURE 4-30 *Visual confinement and safety problems on a bridge with high abutments.*

FIGURE 4-31 *Opening up the view somewhat.*

FIGURE 4-32 *More view and more safety, too.*

are still safety hazards. More recent structures eliminate the side piers and move the abutments back down the slope to a point set by safety clearances and structural economy. Structures began to look like the one shown in Figure 4-32.

The safety problems have now been solved, and the view through the structure is much more extensive. This whole process is an excellent example of the incorporation of aesthetics in structure design. Each step was made for a functional, as well as an aesthetic, reason. At each step, the superstructure cost more, but there were offsetting savings in the substructure. In some cases, these savings totally offset the additional cost; but even when they did not, the new designs were accepted because designers felt that the improved safety and improved appearance were worth the additional cost.

The determining visual variables for abutment placement are girder depth, abutment height (the height of the abutment wall at the bearings), the clearance under the structure, and the ratios between these variables (Figure 4-33). The possibilities vary from massive retaining walls to minimal pedestal abutments perched on the edge of the embankment. The shorter the abutment height, the lighter and less prominent the bridge will appear. The greater the height, the more anchored and heavy the bridge will appear, and views through the bridge will tend to be framed, or, at the extreme, enclosed. On two-span highway overcrossings, the exact position of the abutment may not be a cost issue. Several years ago, the Maryland State Highway Administration did a study that showed that the cost of the additional span length for short abutments was offset by the greater wall and wing wall costs of high abutments.

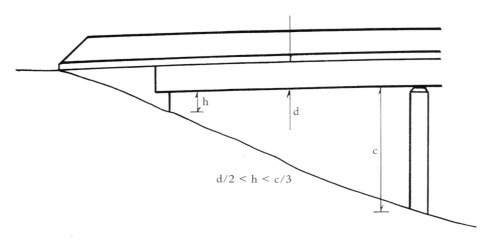

FIGURE 4-33 *The desirable range of abutment heights.*

Relate Abutment Height to the Maximum Height of the First Span

The point at which the abutment becomes a dominant element is when it is greater than one-third the maximum height of the first span. Heights greater than that should be avoided, unless a visually dominant abutment is a desired feature (Figure 4-30).

Figure 2-20 (page 39) shows how a minimal abutment extends the length of the structure and emphasizes its thinness and horizontality. Landscaping at abutments can become a major determinant of the visual impact of the bridge, particularly for small abutments. Abutments of this size can be obscured by landscaping after a few years' growth. Whether this is good or not depends on the designer's overall concept for the bridge. In one- or two-span bridges, hidden abutments will leave the viewer with doubt as to how the bridge is supported (Figure 4-34).

For Structures Where Both Abutments Are Visible at the Same Time, Use the Same Height/Clearance Ratio for Both

A common problem for abutment placement is a structure on a vertical grade, so that one side is higher than the other. Using either a common abutment height or a common distance from the first pier results in an unbalanced appearance (Figure 4-35.1 through 4-35.3).

Use Abutment Wing Walls That Parallel the Upper Roadway

The abutments will usually look best if the wing walls parallel the upper roadway, because this visually "stretches" the structure, making it seem longer and thinner. The end of the wing wall also provides a visually logical place to terminate the approach-roadway guardrail.

Wing walls that simply bisect the angle between upper and lower roadways should be avoided, because they create a major object unrelated to either roadway. They also create triangular areas of landscaping, top and bottom, which are hard to plant and maintain (Figures 4-36 and 4-37). Finally, they require an additional guardrail and a separate parapet end block adjoining the roadway.

FIGURE 4-34 *The effect of abutments hidden by landscaping: The bridge has no visible means of support.*

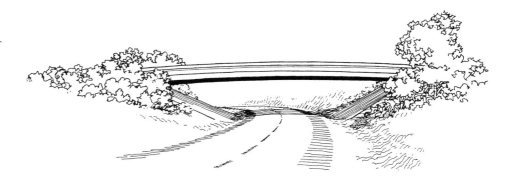

Where Closely Spaced Abutments Are Connected by Retaining Walls, Treat the Abutment Wall as an Extension of the Adjoining Retaining Walls

This situation often occurs in complex interchanges and depressed freeways. The abutment wing walls and retaining walls should blend seamlessly, while the lines and details of the bridge parapets continue along the top of the wall with no breaks (Figure 4-38).

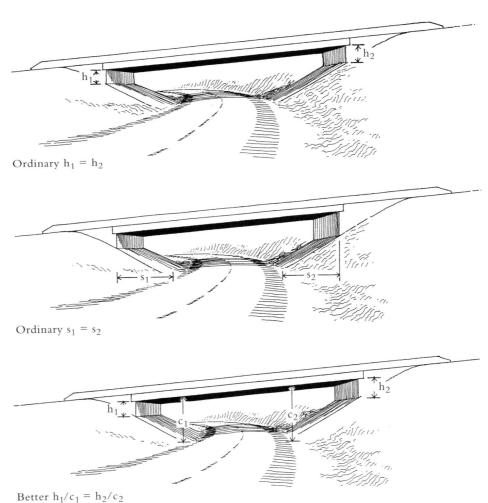

Ordinary $h_1 = h_2$

Ordinary $s_1 = s_2$

FIGURE 4-35.1 THROUGH 4-35.3 *When a structure is on a vertical grade, so that one side is higher than the other, the abutment proportions should be kept similar.*

Better $h_1/c_1 = h_2/c_2$

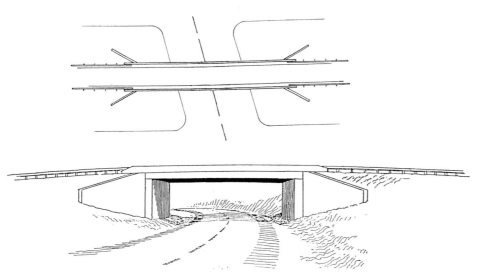

FIGURE 4-36 *When the wing walls are at an angle to both the upper and lower roadways, they become separate elements, which complicate the appearance of the structure; moreover, triangular bits of slope are created, which are hard to maintain.*

For Bridges Longer Than Four Spans, Locate the Abutment in Relation to Nearby Features

For larger bridges, the abutments are a smaller proportion of the structure, and may not even be visible from many key viewpoints. Abutment placement and height will be a function of pier placement and span length throughout the structure, rather than a function of the appearance of the abutments themselves. Abutment placement will have a lesser, though still important, impact on the appearance of the total structure. The more important consideration for longer bridges will be the effect of the abutment on its immediate environs. How well the abutment fits the topography, or matches up with an existing street pattern, will be an important consideration.

If there is an undercrossing roadway near the abutment, then abutment position and height should be determined using the previous guidelines. If the abutment is on a natural hillside, it should be placed to take advantage of a natural promontory or outcropping. In an urban area, it may be appropriate to align the abutment with a building line or setback line. The goal is to find a location that will appear obvious or "natural" in the total setting.

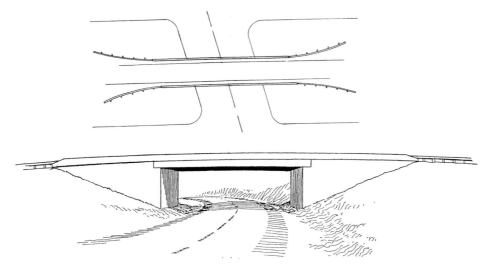

FIGURE 4-37 *Aligning wing walls with the upper roadway simplifies construction, maintenance, and guardrail placement, and improves appearance.*

FIGURE 4-38 *This retaining wall from Albuquerque, New Mexico's Big I interchange is a smooth extension of the abutments it connects with; the wall parapet is a simple continuation of the bridge parapets.*

Abutments for Skewed Structures

With Skewed Structures, Consider Placing the Abutments at the Top of Embankments and at Right Angles to the Overcrossing Roadway

Heavily skewed bridges create difficult abutment problems. If the abutment wall is kept parallel to the lower roadway, it can get quite long, and the slopes of the adjoining embankment can become large enough to become a major visual problem. All of this can be worsened by the large wing walls required to adapt a conventional girder bridge to a skewed condition.

Placing the abutment further up on the embankment and at right angles to the centerline of the overcrossing roadway (Figure 4-39) reduces the visual problem by reducing the amount of fill near the roadway. While it does require longer spans, there are compensating savings in the length and height of abutment walls, as well as significant simplifications in analysis and construction.

The designer must be careful in establishing the height of the abutment on skewed bridges. As the bridge is lengthened, the depth of the structure is drastically affected, sometimes defeating the goal of creating a slender structure, so that structure depth becomes more of a problem than abutment height.

With wider bridges (greater than about 50 feet) and still-lower angles, these ideas no longer suffice.

For Extreme Skews, Align the Wall with the Lower Roadway, Then Flare with a Graceful Curve

Structures on extreme skews can generate wing walls that become quite long and costly. They can, however, be designed to become a positive feature in the scene. When aligned in a smooth curve related to the lower roadway, they help guide the driver through the underpass, reduce the tunnel-like aspects, and provide a sense of transition to the high abutment wall (Figures 4-40.1 and 4-40.2).

FIGURE 4-39 *This right-angled abutment keeps the roadway edge open and reduces the size of abutment walls. The brick on the abutment is a use of traditional masonry material, which is consistent with its historical role and structural capability.*

Retaining Walls

Retaining walls mediate between the existing topography and the roadway geometry. Both of these elements are fixed. As a result, wall height and position are often seen as a mechanical connection of the two, resulting in walls with a haphazard, almost accidental shape. However, there are choices that can be made to improve wall appearance by conscious adjustments in wall position and height (Figure 4-41).

For a Wall between a Lower and an Upper Roadway, Locate the Wall Close to the Upper Roadway

This approach creates space at the edge of the lower roadway, where it will be appreciated by drivers, and increases safety by moving the wall further from the travel way. The upper roadway will require a guardrail anyway, so nothing is lost there. This position will also allow the wall alignment to be independent of the geometry of the lower roadway, which will contribute to the next guideline.

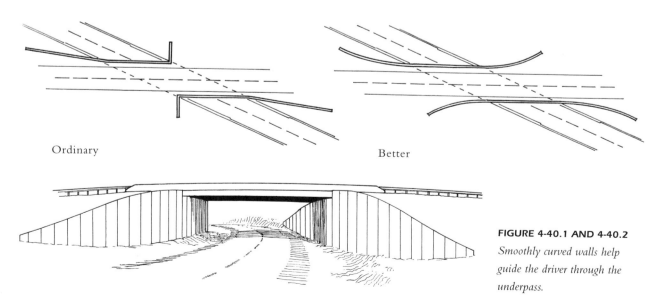

Ordinary

Better

FIGURE 4-40.1 AND 4-40.2
Smoothly curved walls help guide the driver through the underpass.

FIGURE 4-41 *Moving a retaining wall back from lower roadway edge creates more space for travelers and improves appearance while increasing both apparent and actual safety.*

FIGURE 4-42 *The abrupt hump traced by the top of this tiny retaining wall makes it seem arbitrary.*

Align Walls in Continuous Horizontal Curves Related to Roadway Geometry and Topographic Features

Highway environments are made up of continuously curved surfaces. Anything composed of straight edges and angles will seem out of place and even threatening. Walls on curved alignment fit the highway environment; walls aligned on jogs and offsets look out of place.

Shape Wall Tops in Continuous Curves Reflecting and Smoothing Out the Topography

The curvilinear rule applies to wall profiles for the same reason as for plan alignment. A smoothly curved top profile will look like it belongs in the scene. A profile of jumps and bumps won't (Figure 4-42).

Visualize Both Retaining and Noise Walls with Three-Dimensional Drawings

Before finalizing wall designs, develop three-dimensional sketches from the driver's viewpoint showing topographic background as well as the wall. The three-dimensional result of decisions for both retaining walls and noise walls cannot be adequately predicted from viewing plan and elevation drawings.

It may seem like the preceding guidelines would not apply in urban areas. However, since the "topographic features" in urban areas are streets and buildings, applying the guidelines will, in fact, lead to more rectilinear walls that will seem in place in the urban scene, while still recognizing the visual needs of road users.

Noise Walls

Noise walls are usually very large elements in the visual field. Because it is not usually necessary to parallel the highway alignment, designers find themselves with flexibility to follow the vagaries of local drainage or property lines. However, this often results in a jagged alignment, which seems to jerk into and away from the roadway and has an inconsistent relation to the topography (Figure 4-43). A stepped, jagged-top profile can compound the problem.

Many of the rules of retaining walls apply here as well. Control of the horizontal alignment is particularly important.

Align Noise Walls in Long Smooth Curves Related to Major Topographical Features

If the wall is a proprietary wall, which depends on a zigzag alignment for its stability, then the overall alignment of the wall should be established in smooth curves (Figure 4-44).

Such curves can easily be accommodated by the precast concrete panel systems used for many noise walls. Radii as low as 250 feet require a deviation at each post (often spaced about 12 feet apart) of only a few degrees, which is well within the construction tolerances of the precast system.

Profile the Top in Long, Smooth Curves or Small, Even Steps with a Minimum of Height Variation

Because of its position within the visual field, the top of a noise wall is often the most prominent highway element in the scene, particularly if the wall color contrasts with its background. Wall profiles are influenced both by topography and a required acoustical profile. The elements affecting the latter are usually not visible to the driver. Finally, walls are often built from systems of large rectangular panels. Often the result is a wall that proceeds in irregular and apparently irrational vertical steps, with no visual relationship to anything else in the scene. The effect is as disquieting as seeing the jagged edges of torn paper.

The goal should be to shape the top of the wall in some obvious relationship to the topography visible to the driver. Changes in vertical elevation should be made with smooth curves or multiple, identical small steps, rather than a few large jumps.

FIGURE 4-43 *While this brick noise wall has an attractive surface, its zigzag alignment and jagged profile conflict with the smooth curves of the roadway and topography.*

FIGURE 4-44 *A sculptor's approach to placing a large linear form in the landscape: make it flow with the topography. Richard Serra's* Tuhirangi Contour *in Kaukapakapa, New Zealand.*

SUPERSTRUCTURE SHAPE, INCLUDING PARAPETS, RAILINGS, AND BEARINGS

At this point, the superstructure type will have been selected, and pier and abutment locations set. Now is the time to shape the superstructure to reflect the forces on it in ways that improve its appearance. At the same time, choices made now can positively affect the appearance of the pier shapes and abutment shapes to be established later.

The major visual design goals are apparent slenderness, lightness, and continuity, with the structure shaped at each point to efficiently resist the forces upon it. In girder bridges the ideal will give the appearance of a slender, horizontal, perhaps subtly shaped ribbon running from abutment to abutment, and resting lightly on the intermediate piers.

Because the structure is in fact quite heavy and deep, the challenge is to make it *seem* thin and light through a selection of girder depth, shape, and details. For example, the haunches and small exposed bearings of the U.S. Naval Academy Bridge in Figure 1-22 (page 19) create the illusion that the main span is resting on four points, and therefore must be very light indeed. The design of the cross section and, in particular, the edge profile, has a major impact on the overall appearance of the bridge because the parapet and fascia girder are often the most visually prominent parts of the bridge and influence the perception of the structure's depth. This is especially true for highway overcrossings, because parapets are such a prominent part of the structure's elevation view.

Girder Shape and Depth

The major elements to be settled are the superstructure depth, shape, and slab overhang, as they will have a powerful effect on appearance. The overhang, along with the parapet face, provide the strongest opportunity to make the bridge seem more slender than it really is. Two relevant guidelines are to minimize girder depth and to maximize the overhang, which are, to some degree, contradictory. A balance must be sought between them that will result in the best combination of performance, cost, and appearance.

> *Seek Girder Arrangements That Emphasize Apparent Thinness and Horizontality*

In the typical highway environment, the horizontal is the dominant dimension. As noted in Chapter Two, most people will find most attractive the structure that appears the thinnest and with the most horizontal emphasis.

The techniques available are:

- Maximize the overhang; maximize girder spacing. Maximizing the overhang creates a strong shadow on the fascia girder, which visually divides the girder horizontally (taking advantage of one of the visual illusions shown in Figure 2-16, page 37) or places the girder entirely in shadow. The shadow thus reduces the apparent depth of the fascia girder. This can be a particularly important feature in long single-span structures. Maximizing the overhang

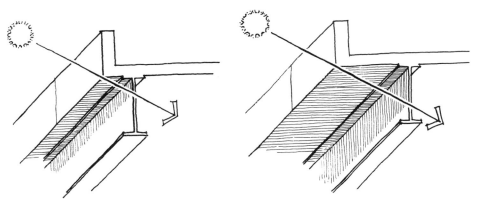

FIGURE 4-45 *Larger overhangs create a larger shadow line.*

pays dividends in the substructure as well, particularly for skewed structures (Figure 4-45). The pier caps and abutment end walls can be shorter, and fewer columns per pier will be required. This means that oblique views through the structure are less likely to look like a forest of columns. The amount of the overhang will also allow more light to enter the underbridge area, because the fascia girder is farther back from the structure's edge (Figure 4-46).

The need to have scupper outlets inside the fascia girder is often given as a reason to limit the size of overhangs. There are several solutions to this problem, which are discussed in the Chapter Five.

Maximizing girder spacing allows a larger overhang without varying slab depths. With I-girders, a limit will be reached for narrow ramps, as a minimum of three girders is required for redundancy. Box girders have more internal redundancy and torsional stiffness and can more easily accommodate large overhangs. The drawback for this guideline is that it requires that more load be carried by each girder, which may require a deeper girder, thereby somewhat

FIGURE 4-46 *Box girders can more easily accommodate large overhangs.*

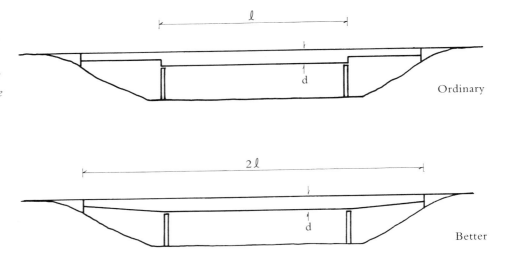

FIGURE 4-47 *Visual continuity makes the bridge seem thinner because the whole bridge is seen as one long unit, while the maximum girder depth stays the same.*

offsetting the effect of the deeper shadow line. Which combination of depth and overhang is best must be determined for each structure. The following is a useful target.

- Have the minimum overhang equal the girder depth. This proportion will place the girder, or a significant part of it, in shadow most of the time, thereby minimizing the apparent depth of the structure. Figure 2-20 (page 39) shows the effectiveness of this approach.

Keep Structural Depth Constant or Smoothly Varied over the Bridge

People observe bridges by sweeping their eyes along the length of the structure, therefore any abrupt changes in girder depth will be jarringly obvious. Constant or smoothly varying depth will make the structure more unified. It will also make the structure *appear* longer, and therefore thinner. The eye judges relative thinness by comparing length to depth, so making the girder seem longer will also make it seem more slender. Structural continuity is a big help here, as it keeps everything visually continuous as well.

With girder bridges, the spans will often vary in length over the bridge. It will often be necessary to change girder depth, provide haunches, or do both in a single bridge. Transitions between girder depths should be made with girders that are tapered or have a curved lower flange (Figure 4-47).

Consider Haunched Girders Where Feasible

Haunches are important because they visually demonstrate the concentration of forces in the bridge and make the bridge seem thinner by reducing the average depth while leaving the length the same. Haunches provide an important point of visual interest as well. If haunches are not feasible, the structure will depend for its interest primarily on the proportions of the girder, particularly its relative thinness, as compared to the other features of the structure.

- Make haunches long enough to be in proportion to the span length (Figure 4-48). The moment diagram is a good guide (i.e., start the haunch at the point of inflection). In order to reduce fabrication costs of steel girders, the splice

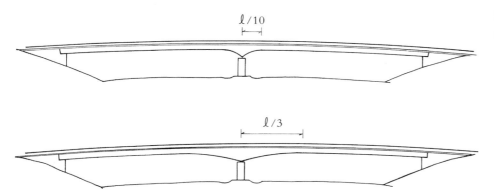

FIGURE 4-48 *The effect of short versus long haunches.*

point can be set at the point the haunch begins, and the haunch depth can be set so that flange plate thickness changes only at the splice point, which eliminates the need to butt weld flange plates.

- Use pointed haunches because they concentrate visual interest at the point of force transfer and better reflect the flow of forces. "Fishbelly" haunches make the girder look heavier and waste material (Figure 4-49).

- Haunches are usually better when formed by smooth curves. Parabolic curves work well. Straight lines can make the haunch seem shorter if the break point between the haunch and the rest of the girder is too obvious; then the girder appears to be divided at that point (Figure 4-50). If using straight lines, make the angle at the beginning of the haunch no greater than about 8 degrees.

- Don't make haunches deeper than about twice the mid-span depth. Otherwise, the mid-span will look too fragile and the haunch will look too heavy.

- Conversely, don't make the depth at the haunch too close to the mid-span depth or the haunch will be imperceptible (Figure 4-51). A reasonable minimum is 1.3 times the mid-span depth (see Figure 2-16, page 37 for an example of too small a difference). The angle at the point of the haunch should be

FIGURE 4-49 *Visual (and structural) weight added by a "fishbelly" haunch.*

FIGURE 4-50 *Forming a haunch with straight lines and a short break makes the haunch seem short.*

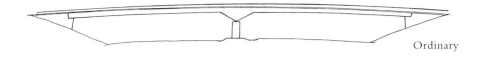

Ordinary

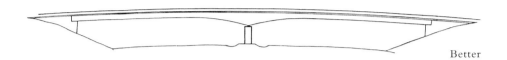

Better

FIGURE 4-50 *Forming a haunch with straight lines and a short break makes the haunch seem short.*

between 135 and 160 degrees; otherwise, the bearing point will seem too delicate (Figure 4-52 and 4-53).

Consider Use of a Haunch with Skewed Structures

Skew structures can create illusions about span length and depth. The nearer span will seem longer and deeper than the balance of the bridge, often resulting in a distorted appearance (see Figure 4-27.2, page 116). A haunch may correct this problem. Reshaping a skewed girder to add a haunch makes it appear thinner and compensates to some degree for the distortion caused by the skewed alignment. Compare Figure 1-23.1 (page 22) to Figure 1-23.2 (page 22).

Shapes of Arches and Frames

For Arches, Shape the Arch Rib to Reflect the Flow of Forces, with the Thinnest Section at the Hinges

Arches are also inherently interesting forms to begin with. Reducing the thickness in areas of reduced moment will make the structure even more attractive. Salginatobel owes its fame to Maillart's application of this simple idea (Figure 1-18, page 13, and Figure 4-54).

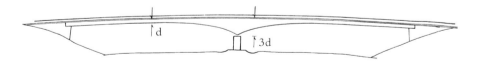

d   3d

FIGURE 4-51 *Too much depth at the haunch makes the girder look heavy.*

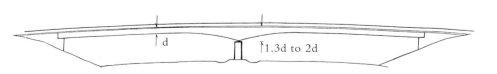

d   1.3d to 2d

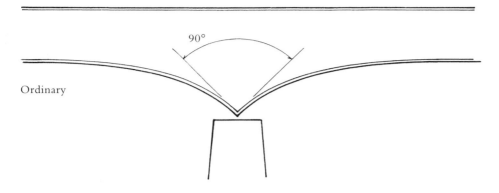

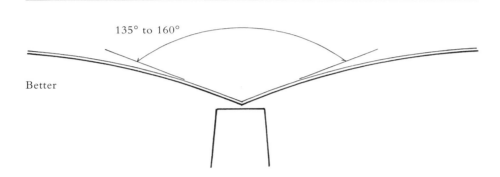

FIGURE 4-52 *The desirable angle at the bearing point.*

Keep Arch Deck Supports (Spandrel Columns) Similar in Thickness over Bridge's Entire Length

With regard to the deck supports, a heavier support at the arch abutment is usually structurally unnecessary and visually interrupts the lines of the bridge. Figure 3-21 (page 83) shows the Rainbow Bridge in Folsom, California, an arch with a heavy pier at the arch abutment. The bridge would seem much more continuous with the approaches without these elements.

FIGURE 4-53 *Haunches demonstrate the flow of forces and make this Colorado bridge seem thinner.*

FIGURE 4-54 *The visual drama of Maillart's Salginatobel Bridge is the result of a structural decision to reduce the amount of material where it is not needed, at the hinges.*

With Rigid Frames, Give the Legs Enough Slant to Maximize Their Lengths and to Give Full Play to the Visual Illusion of Additional Length

Frames are inherently interesting shapes to start with. Shaping the frame to respond to the areas of minimum moment enhances that interest. Shaping rigid frames and delta frames takes experimentation. Much depends on the relative lengths, vertical clearance, and shape of the side slopes. Written guidelines can be misleading; better to make sketches of multiple shapes, as seen from the important viewpoints, than to pick the one that *appears* most graceful.

The leg width at the joint with the girder should approximate the depth of the girder, and both girder and legs should taper to a minimum at the supports. Figures 2-37 (page 50) and Figure 4-21 (page 112) show nicely proportioned rigid frames.

For Both Frames and Arches, Keep Sway Bracing and Flooring System to Simply Arranged Minimum Number of Members and in a Clear, Consistent Relationship to the Main Members

No matter how graceful a frame or arch looks in an elevation drawing, the three-dimensional real-life view can be significantly degraded by a confusion of secondary bracing members.

The bridge in Figure 6-26.2 (page 217) recognizes and controls this potential problem.

Parapets

Parapets and railings serve functions for people both on and below the bridge. And because each side has its own requirements, the two have to be reconciled during the design process.

The first issue to be faced is whether any of the parapet can be open or must it be entirely solid. The overall superiority of the Jersey barrier from a safety point of view has settled this issue in most cases in favor of a solid barrier parapet. However, this forecloses much of the driver's view from the bridge; it also restricts the options available in using the parapet to influence the appearance of the bridge elevation. Many open bridge rail designs have now been developed, and their safety character-istics have been proven through crash testing. They should be considered as an option to the typical concrete Jersey barrier. The discussion that follows starts with the solid part of the deck edge. Railings are discussed in the next section.

The parapet is an important influence on the overall appearance of a bridge because it—and the girder—determine the visual depth/span ratio of the super-structure. This ratio is one of the strongest influences on appearance.

One key dimension is the height of the parapet relative to the exposed depth of the girder. For this dimension, the German engineer Fritz Leonhardt[8] suggests the following guideline.

Keep the Parapet Height between One-Fourth and One-Half of the Exposed Girder Depth with a Minimum of One-Eightieth of Span Length

For most medium-span bridges, the standard Jersey barrier parapet will be higher than this ideal (Figure 4-55). Attention should then be turned to ways of

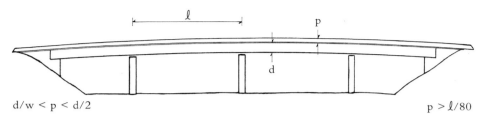

FIGURE 4-55 *The visually desirable parapet height.*[8]

d/w < p < d/2   p > ℓ/80

making the barrier appear thinner by application of one of the visual illusions discussed in Chapter Two.

Break Up the Face of the Parapet Horizontally, Using Incisions, Recesses, or Sloped Planes

The apparent depth of the parapet itself is also subject to manipulation through the same type of visual illusion. The available techniques are:

- Divide the parapet fascia into separate horizontal surfaces by shadow and/or physical breaks. Elements that create apparent horizontal division in the combined girder–parapet will make the bridge seem longer and the combined girder–parapet seem thinner. The relative size of these divisions also makes a difference. Evenly spaced divisions do not seem as effective or attractive as uneven divisions, where one horizontal element is thicker than the next. (Figure 4-56, left).

- Change the relative brightness of different fascia surfaces by changing their angle so that they catch more sun, or no sun at all (Figure 4-56, center).

- Remove the horizontal divisions between fascia surfaces by introducing curved connecting surfaces, which leave the viewer fewer clues by which to judge thickness (Figure 4-56, right).

Any Parapet Pattern Should Have Dominant Lines That Are Horizontal

Horizontal details should be emphasized in the selection of patterns for the parapet (Figure 4-57). Vertical details will interrupt the dominant lines of the bridge and make the parapet look deeper. Spaces between vertical divisions in the parapet should be at least three times parapet depth. There is an exception for vertical patterns that are so closely spaced that they read as textures rather than distinct vertical lines. Such patterns tend to hide drainage stains and construction irregularities. Form liners that create closely spaced vertical projections ("fins" or, if irregular, "fractured fins") have

FIGURE 4-56 *The effects of cross-section differences on appearance.*

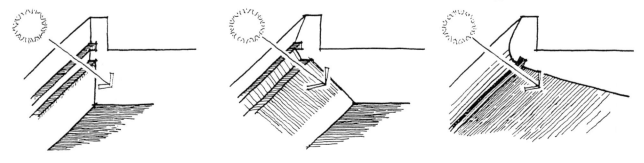

been used to create these textures, and can even be slip-formed, as on the bridge in Figure 2-55, page 57.

Recognize the Slab/Parapet Joint with a Significant Groove or Recess Incorporated in the Parapet Design

The choice of construction technique will have an important influence on parapet details. The best results can be obtained when the parapet covers the slab edge. This allows for any irregularities in the slab edge to be eliminated and produces the best control of alignment. It also eliminates staining the parapet face by roadway drainage seeping through cracks at the slab/parapet joint.

That said, the relatively greater difficulty of forming the element covering the slab edge has led some agencies to leave the slab edge exposed. If it will be exposed, the slab/parapet construction joint must be recognized; it is impossible to effectively camouflage it (Figure 4-58). The construction joint should be incorporated into parapet patterns aimed at making the parapet look thinner.

Precast parapets have been experimented with; however, they have proven to be difficult to align accurately in the field. Precast, stay-in-place forms for the parapet avoid some of these problems and provide many options for shape, color, and finish; can be produced to higher tolerances and a denser surface than field-place concrete;

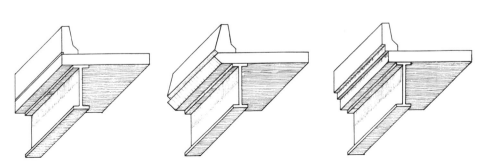

FIGURE 4-58 *Incorporating the slab parapet joint in the parapet design.*

and are easier to align. Precast panels attached to poured-in-place parapets are another attractive option.

When Designing the Parapet Face, Use Details Large Enough (4 Inches Minimum) to Be Recognized at Highway Distances

Parapet details have a particularly strong impact on highway overcrossings, as the parapet and fascia girder are usually the most visible parts of the structure. Any parapet features must therefore be large enough to be seen and appreciated at highway distances and speeds.

Use Shapes on the Parapet That Are Consistent with Either the Girder Cross Section or the Pier Shapes or Both

Precast I-girders have characteristic faceted edges that should be emulated with faceted shapes on the parapet (Figure 4-59). Steel wide-flange or welded I-girders and box girders have thinner, sharp-edged shapes that can be emulated in the parapet details. Or, because the visual characteristics of steel girder cross sections are not particularly distinctive, the cue can be taken from a memorable pier shape.

Check Parapet Profiles at Points of Superelevation Transition Using Profile Drawings with Exaggerated Vertical Scales

Because the top of the parapet is a nearly horizontal line parallel to the line of sight of a driver on the bridge, any flaws in the alignment, whether due to design or construction, will be magnified. Problems most often crop up at transitions of superelevation, because the parapet profile is driven by pavement and shoulder cross-slopes.

Prevent Drainage Stains on the Parapet from Railings and Fences

Because of the prominence of the parapet, drainage stains have a significant affect on the bridge's appearance. These result primarily from dirt, de-icing salts, and corrosion products from the fence or railing, and are worst at the railing posts. One or more of the following techniques should be applied:

- Slope the parapet top slightly to drain toward the roadway side of the parapet.

- Provide incisions on the parapet face under railing posts to control drainage streaks.

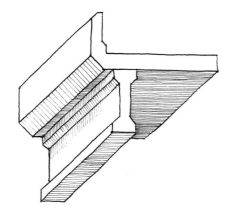 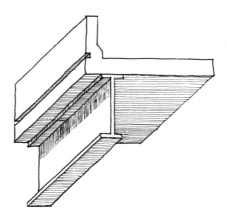

FIGURE 4-59 *With precast girders, the parapet shape should be from the same family of shapes as the girders.*

- Use corrosion-proof fence and railing materials.

- Coat the parapet with a stain-resistant finish.

- Provide vertical striations in the parapet to hide stains.

End Parapets, Railings, and Pedestrian Screens by Tapering Them Down or Flaring Them Away from the Traveled Way, or Doing Both; Do Not Use Vertical End Blocks

At each end of the bridge, the parapet has the visual job of gracefully ending or beginning the bridge, and the functional job of providing a place for the approach guardrail and right-of-way fence to tie in (Figures 4-60.1 and 4-60.2). The goal is to make the transition as smooth and continuous as possible, with a minimum number of separate elements. Aligning the abutment wing wall with the upper roadway helps by allowing the parapet end block and wing wall to be combined.

Incorporate Attachment Details for the Guardrail and Right-of-Way Fence into the End of the Parapet

This approach fits the overall goal to emphasize the horizontality of the bridge and its continuity with the approach roadway. It also has safety benefits.

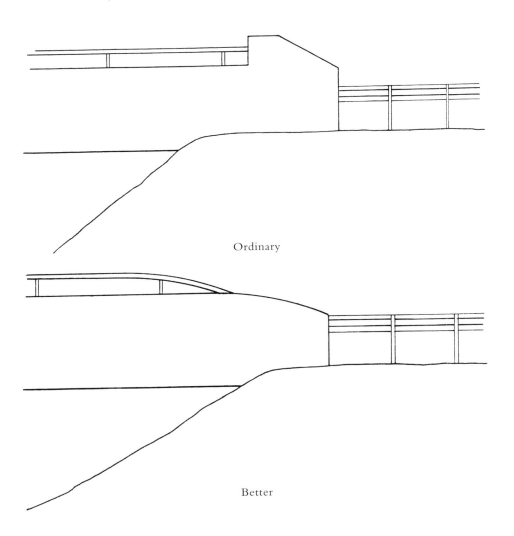

Ordinary

Better

FIGURE 4-60.1 AND 4-60.2

A tapered ending simplifies the parapet ending while accommodating guardrail and other features.

Railings, Pedestrian Screens, and Noise Walls

Railings, pedestrian screens, and noise walls will be seen together with the parapet and the structure itself. All of these elements need to be considered simultaneously.

Railings can be positive features on a bridge, particularly if they substitute for all or part of a solid parapet. Open railings are greatly appreciated by bridge users, particularly in scenic areas. A pedestrian screen or noise wall on a bridge complicates matters because it can make the parapet seem three times as high, with comparable negative effects on the proportions of the whole structure.

Major Horizontal Railing and Screen Members Should Be Significantly Larger Than Vertical Members and Not Be Interrupted at the Vertical Members

As with the parapet itself, the details of railings, pedestrian screens, and noise walls should maintain a horizontal emphasis.

Place Posts in Some Consistent Relationship with Contraction Joints and Other Visible Features of the Parapet

If there is any visible pattern on the parapet, even if it is just a series of concrete contraction joints, the posts need to have a consistent relationship with that pattern (Figure 4-61).

Use Railing Details and Other Details That Complement the Shapes of Major Structural Features

The structure will appear more unified if the details are clearly similar in shape to the major features. (See Figure 5-56, page 184.)

Recognize That Curved or Slanted Posts Contribute to the Horizontal Emphasis by Eliminating All Vertical Lines

Slanted or curved supports put the horizontal members in the dominant position.

Use Materials, Details, and Colors for Pedestrian Screens That Are as Transparent as Possible and That Make Clear That the Screen Is Separate from the Parapet

The goal is to keep the screen from increasing the apparent depth of the bridge. This can be done by making the screen a visually different element from the rest of the bridge. Color, pattern, and texture are all tools to achieve this visual difference. For example, the screen should have a clearly different color from that on which the

FIGURE 4-61 *The horizontal members of screen support and railing systems should dominate; the relationship of posts to parapet divisions should be consistent.*

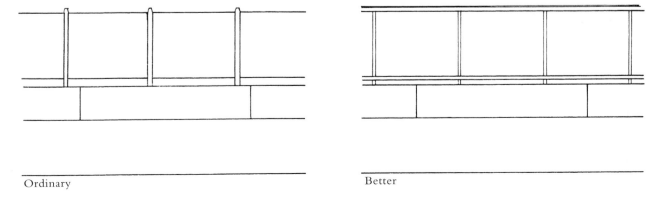

Ordinary   Better

FIGURE 4-62 *A simple strategy of pairing the railing posts and connecting them with a perforated plate gives this pedestrian screen a memorable appearance. The posts stay below the top rail, so the horizontals remain dominant. The coordination of the plate design with the triangular parapet pattern ties the two together. The parapet pattern is produced by alternating sections of smooth concrete with sections that have a closely spaced vertical rib texture.*

parapet on which it is mounted. Pedestrian screens are often seen against the sky. Colors such as light blue-gray or gray will tend to take on the sky color and seem to disappear (Figure 4-62). Alternatively, pedestrian screens seen against wooded or rocky backgrounds can take their cue from those backgrounds.

Clamps, Elbows, Junctions, and Other Fittings Should Be Compact, Simple, and Rustproof, and Integrated with the Major Members Wherever Possible

The goal here is to keep the appearance of the railing or screen as simple as possible and to prevent these functional but secondary elements from distracting from the appearance of the bridge or the railing/screen itself.

Design Noise Barriers with a Pattern or Color to Visually Separate the Barrier from the Bridge

Noise barriers must be solid. However, as with screens and railings, the goal is to keep the barrier from increasing the apparent depth of the bridge. The barrier can be made to seem separate from the structure by differences in color, pattern, and/or texture. Some noise barriers on structures have been successful using specific pictorial or abstract designs on the community side.

FIGURE 4-63 *A bearing compared to a moment connection.*

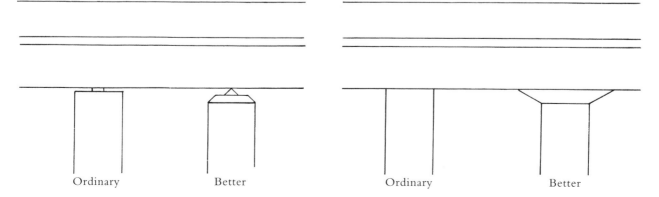

Ordinary   Better   Ordinary   Better

At each end of the bridge a graceful ending of the noise barrier is necessary. Since the overall goal is to emphasize the horizontality of the bridge and its continuity with the approach roadway, the barrier alignment should curve smoothly away from the bridge to meet the barrier alignment on the nearby roadway. The top profile should also be smoothly curved.

Bearings and Stiffeners

The point at which the girder connects to the pier is crucial visually as well as structurally. The rule is simple: Only material necessary to resist forces should be present, leading to a different appearance at a bearing than at a moment connection. Elements needed anyway for structural reasons (stiffeners, bearing plates, etc.) can be designed as attractive details that accentuate the points of stress concentration and transfer.

Use High Bearings, Pedestals, or Chamfers at Pier Tops to Visually Attenuate the Bearing Point

This approach elongates the bearing or raises the bearing on a small pedestal so that it is silhouetted against the sky, visually separating the superstructure from the pier with what appears to be a tiny element. This makes the bridge seem light in weight, in the same way that a waiter makes carrying a heavy tray seem effortless by balancing it on his fingertips. This approach also avoids any visual interruption of the horizontal lines of the superstructure, which makes the structure seem longer and more slender. Figure 3-2 (page 62) shows one way this can be accomplished. The bridges in Figure 4-4 (page 100) and 5-25 (page 156) show another way. Compare Figure 4-4 with Figure 5-20.1 (page 154) to see the difference between bearing and moment connections. In no case should pilasters be used to hide the bearing.

For Moment Connections, Thicken and Round the Joint Area to Accommodate the Stress Concentrations That Occur

Based on their familiarity with moment connections in their own experience (their own shoulders and hips, tree branches), people recognize the point of stress concentration, expect that there will be more material there, and are disappointed when the shape of the structure does not match that expectation. Figures 6-32 (page 226) and 5-20.1 (page 154) are good examples of appropriate moment connections.

The spacing of stiffeners on a girder, or the intersection of flanges at a truss connection, create patterns that will be recognized. Because of the prominence of the fascia surfaces, the stiffeners on steel-plate fascia girders are especially significant visual features. They will create an impression, which should be controlled.

Design the Pattern of Stiffeners on Steel Fascia Girders to Accentuate the Point-of-Stress Transfer

Vertical stiffeners are usually necessary at bearing points, where they confirm people's intuition that something important is happening there. Using two or three stiffeners, or arranging multiple stiffeners in a V-shape can create a strong visual effect at very little cost (Figure 4-64).

FIGURE 4-64 *The sloped web stiffeners focus attention on the point of force transfer.*

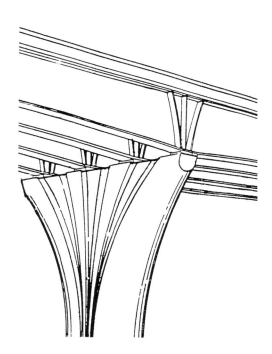

Avoid Evenly Spaced Vertical Stiffeners on the Exterior of Fascia Girders

Evenly spaced vertical stiffeners will divide the girder vertically and make it appear thicker (Figure 4-65). If vertical stiffeners are necessary on fascia girders, they should be placed on the interior face of the girder. There are two exceptions to this guideline:

- If the stiffener spacing exceeds the girder depth by at least a factor of 2, the resulting rectangles have a dominant horizontal dimension, and the visual problem is much reduced.

- If the spacing of vertical stiffeners is allowed to vary continuously in direct proportion to the shear stress the resulting rhythmic pattern can be visually strong enough to override the thickening effect. The bridge shown in Figure 6-75 (page 269) has some spacing variations, but does not carry the idea far enough to create a visually convincing result.

Consider Horizontal Stiffeners Because of Their Ability to Reduce the Apparent Depth of the Deep Girders

Horizontal stiffeners on the outside of fascia girders have a positive visual effect: They subdivide the girder horizontally and make it appear thinner. If the girder is

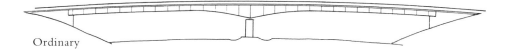

Ordinary

FIGURE 4-65 *Vertical stiffeners will make the girder seem heavier than a comparable girder with horizontal stiffeners.*

Better

haunched, a partial horizontal stiffener over the support curved to a pattern complementary to the lower flange will reinforce the visual effect of the haunch. Engineers should be aware that current fatigue requirements specify careful design of horizontal stiffeners in the tension area.

1   This notion was introduced by Christopher Alexander in the following books on architecture: Alexander, Christopher, 1979. *The Timeless Way of Building.* New York: Oxford University Press; and Alexander, Christopher; Ishikawa, Sara; Silverstein, Murray; Jacobson, Max; Fiksdahl-King, Ingrid; and Angel, Shlomo, 1977. *A Pattern Language: Towns-Buildings-Construction.* New York: Oxford University Press.

2   The author's firm, Rosales Gottemoeller & Associates, was aesthetic advisor to JMI International (now EarthTech) for the design of the Martin Luther King Parkway Interchange and the Raccoon River Bridges.

3   This term was originated by Tunnard and Pushkarev in *Man Made America, Chaos or Control.* More detailed discussion of these ideas can be found there.

4   The author's firm, Rosales Gottemoeller & Associates, was aesthetic advisor to Michael Baker Jr. Inc. for the design of this project, the Route 52 Causeway Replacement between Somers Point and Ocean City, New Jersey.

5   Price, Ken, 1993. "Economical Steel Box Girder Bridges," *Transportation Research Record 1393 Structures.* Washington, DC: Transportation Research Board.

6   Azizinamini, A, Van Ooyen, K., Jabar, F., and Fallaha, S., 2003. "High-Performance Steel Bridges: Evolution in Nebraska," 2003 Lecture Series, Illinois Section, American Society of Civil Engineers, Chicago, IL.

7   American Segmental Bridge Institute and Precast/Prestressed Concrete Institute, 2002, *Segmental Box Girder Standards for Span-by-Span and Balanced Cantilever Construction, Edition II,* American Association of State Highway and Transportation Officials, Washington, D.C.

8   These recommendations are from Leonhardt, Fritz, 1982. *Brucken.* Cambridge, MA: The MIT Press.

chapter five

A DESIGN LANGUAGE: GUIDELINES FOR THE SECOND FIVE DETERMINANTS OF APPEARANCE

"The next note must seem fresh but inevitable."

—LEONARD BERNSTEIN, AMERICAN COMPOSER AND CONDUCTOR

This chapter looks at the second five Determinants of Appearance: pier shape; abutment shape; color; surface texture and ornamentation; and signing, lighting, and landscaping. Once the first five determinants have been positively applied, decisions about these elements can add to the interest and success of the bridge.

PIER SHAPE

Piers can be a major element in forming people's impression of a bridge. This is particularly true of girder bridges, where the superstructure does not present as much opportunity to create a memorable image. There is no single correct type of pier; what is appropriate and good-looking on a narrow ramp overpass will be different from what is appropriate and good-looking on a wide dual structure. There should be a clear visual relationship between the substructure elements, meaning that potential abutment features need to be kept in mind when developing pier shapes.

The appearance of piers is heavily influenced by their proportion: their length relative to their height. As piers get taller, the engineering challenge may increase, but the aesthetic challenge decreases. In other words, it is easier to make a tall pier look good than a short pier. In order to highlight the differences, the guidelines in this section are divided into two categories of piers: short piers, those that are wider than they are high, and tall piers, those that are higher than they are wide (Figure 5-1). Obviously, drawing this line between the two categories is arbitrary, and there is a continuum of shape involved. However, there are enough definite differences in the aesthetic problems involved with the two types of piers to make it worthwhile to analyze them in this way.

FIGURE 5-1 *Short and tall*
piers defined.

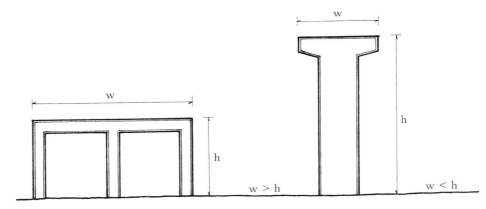

FIGURE 5-1 *Short and tall piers defined.*

Longer multi-span structures present particular difficulties with pier shape. They may have piers of significantly different height and/or width, some of which fall in the short category and some in the tall category, and all of which have to be accommodated within the same family of shapes.

Short Piers

Why are short piers so difficult? Because the pier cap is such a large portion of the total pier. Thus, the pier cap introduces a third element into the visual scene. The overcrossing roadway/superstructure is one element; the columns with their vertical lines are another. The pier cap is clearly not part of the superstructure, but neither is it clearly part of the columns, especially if the columns are geometrically separate shapes. The mind and the eye have a hard time dealing with the complication introduced by this third element, hence it becomes a distraction.

The end of the pier cap compounds the distraction. The end of the pier cap is usually at about the same plane as the parapet fascia and fascia girder. It generally extends out from under the shadow of the overhang and is a relatively bright surface (Figure 5-2). Visually, the end of the pier cap "reads" as part of the superstructure, which interrupts the lines of the superstructure and makes the superstructure appear deeper.

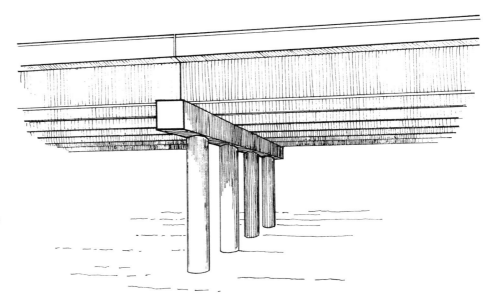

FIGURE 5-2 *The prominence of the pier cap and especially its end surface. The pier cap end visually attaches to the superstructure and makes the bridge seem thicker.*

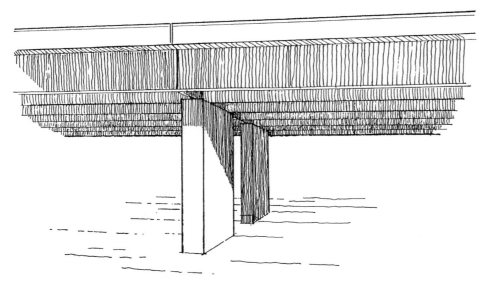

FIGURE 5-3 *Integrating the pier cap into the superstructure removes it as a visual element.*

Use Piers That Eliminate or Minimize the Pier Cap

The key to improving the appearance of a short pier is to eliminate or minimize the pier cap, or incorporate it into the superstructure.

- Use pier types with no cap. Enhanced diaphragm framing can be used to combine girder reactions over a narrow column, or the pier cap can be integrated as a cross girder into the superstructure (Figure 5-3; also see Figure 4-14, page 108, and 5-20.2, page 154). Another approach is to use a solid shaft or wall for support (Figure 5-4). The wall can be made relatively thin, so this can result in an elegant structure. The leading and trailing edges of the wall can be shaped to add interest. When the wall gets long because of overcrossing width or skew, the lack of visibility through it will be a problem. Also, do not use solid (wall) piers as side piers, as they cut off the view completely.

- Recess the pier cap behind the front column, or otherwise unify the pier cap and columns so as to minimize the pier cap as a separate element. One varia-

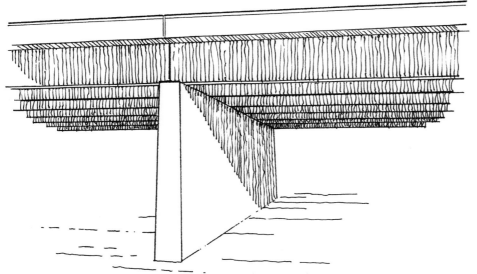

FIGURE 5-4 *Simple wall pier means no pier cap or pier cap end.*

FIGURE 5-5 *Pier cap recessed behind front column.*

tion of this option is to make the cap thinner than the columns (Figure 5-5). Another is to unify the cap and columns by design into a single continuous form (Figure 5-6). These options eliminate the end cantilever. They may not, however, be appropriate in situations involving a family of adjacent piers, some of which are hammerheads.

- Minimize the end elevation of the pier cap by keeping the vertical dimension significantly smaller than the horizontal. Achieve this by: tapering the pier cap cantilever in one or two dimensions, sloping all or part of the end surface, sloping the pier cap sides so that the end elevation is a keystone shape, or using some combination of these techniques. The visual effect of these techniques is to make the cap seem more a part of the girder (Figure 5-7).

For Skewed Structures, Use One Column Per Girder or Girder Pair

This eliminates the pier cap as an element. Because of the additional pier length on skewed bridges, and because the far-side piers of the skewed structure are more visible than in the typical highway overcrossings, pier design becomes more important on skewed bridges. At the same time, it becomes more difficult because the pier

FIGURE 5-6 *For simple structures, the simplest solution is best as shown in this bridge with no pier cap end.*

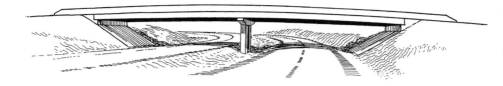

FIGURE 5-7 *A minimal pier cap end; dominant end dimension is horizontal.*

cap gets longer and more prominent. If one column is used for each girder, the need for a pier cap disappears (Figure 5-8). This is a particularly good strategy for box girders. A variation that is particularly effective for skewed bridges is to use individual rectangular columns aligned with the bridge, which creates a "venetian blind" effect on the undercrossing roadway.

Limit Taper of V-Shaped Piers

V-shaped piers eliminate the pier cap. However, when short, they create another visual problem because they are counterintuitive: They are thinnest where visual (and structural) logic says they should be thickest—at the bottom. Tapers that are too extreme produce piers that look top-heavy and illogical. The viewer will ask, "Why not just carry the pier right down to the ground?" (Figure 5-10, top). Or, a taper that looks reasonable on a short pier will produce a pier that is too wide at the top when applied to a taller pier (Figure 5-11). Finally, when the bottom width is too thin compared to the total width, these piers will look unstable, as if they were about to topple over (Figure 5-12). These problems can be alleviated by a proper degree of taper, which controls the proportions of the total width to the width at the narrowest point.

- Keep taper to slopes of 1:4 or less for piers tapered lengthwise; 1:16 or less or piers tapered across their width. Use less for taller piers.

- Make the base width at least one-third of pier width at the top.

Limit Taper of A-Shaped Piers

Piers tapered in the other direction, narrow at the top and wide at the bottom, can suffer from similar problems in reverse. A taper that looks reasonable on a short

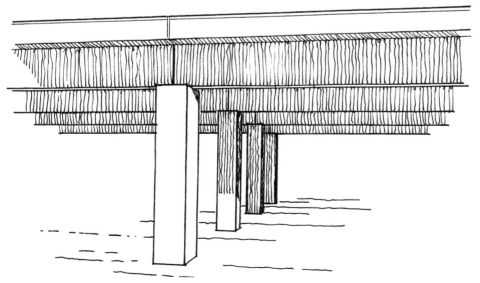

FIGURE 5-8 *With one column per girder, there is no need for a pier cap.*

FIGURE 5-9 *The torsional stiffness of box girders allows wide overhangs, which minimizes the required width of the pier shaft. The V-shape for the overall pier, combined with tapered arms of the V, make this a graceful pier that gives an open and transparent appearance to the bridge and makes the superstructure seem light in weight.*

pier will produce a pier that is too wide at the bottom when applied to a taller pier. The pier will start to look like a pyramid. Use the same slopes as suggested above.

Form Hammerhead Piers into Structurally Logical Shapes

Short hammerhead piers produce similar problems of proportion. While it is difficult to give specific rules, shapes that are structurally logical will also generally be visually logical (Figure 5-13). The key areas of concern are:

- Cantilever length compared to pier height. A cantilever that is longer than the distance between its underside and the ground will appear unnecessary and wasteful.

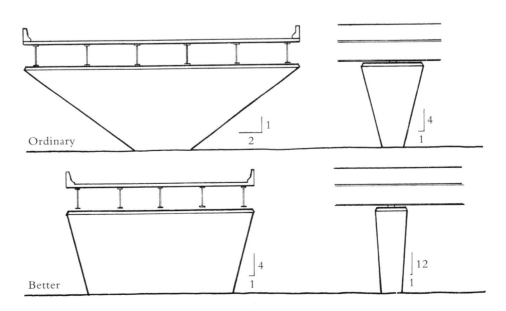

FIGURE 5-10 *V-shaped piers with too much taper will look top-heavy.*

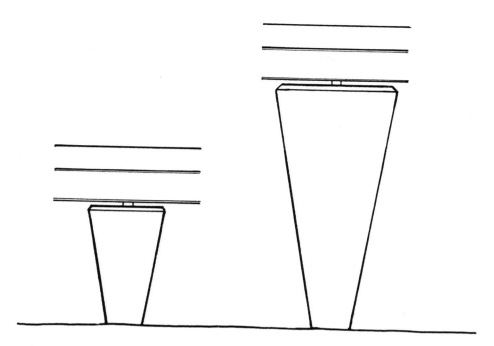

FIGURE 5-11 *Using the same taper on piers of different heights will make the taller piers appear top-heavy.*

- Pier width compared to top width. If base width approaches top width, the cantilevers will be so short the viewer will wonder, "Why did the designer bother to add them?"

- Cantilever depth at the shaft compared to arm length. Thin, long arms will look fragile.

- Cantilever depth at the shaft compared to depth at the tip. Cantilevers should be noticeably deeper at their support than at their tip. Based on their familiarity with cantilevers of their own experience (their own arms and legs, tree branches) people expect a cantilever to be thinner at the end, and are disappointed when the shape of the structure does not match that expectation. Of course, a tapered cantilever is also more efficient structurally.

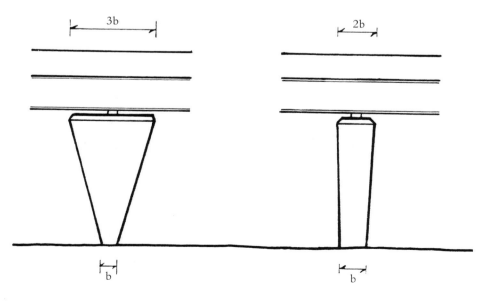

FIGURE 5-12 *V-shaped piers that are too narrow at the base will look unstable.*

FIGURE 5-13 *Do's and don'ts of short hammerhead piers.*

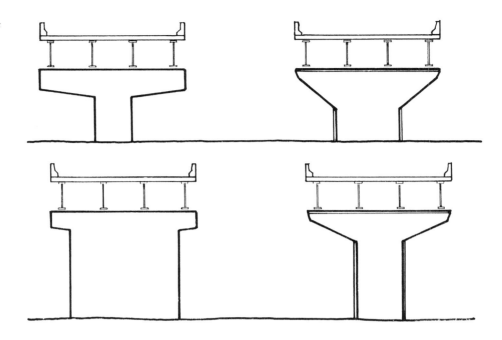

Minimize Cantilever Ends Using the Same Techniques as for Pier Cap Ends

The cantilever ends will create the same visual problem as discussed previously for pier cap ends.

Make Pier Width Proportional to Superstructure Depth, Span Lengths, and Visible Pier Height

On short piers, width should be considered for its visual effect. (On tall piers, structural and economic issues tend to make this decision.) Piers noticeably thicker than the depth of the superstructure will look heavy and squat (Figure 5-14). Pier proportions should recognize that an open railing significantly changes the apparent thickness of the superstructure. Use of a mound for pier protection will shorten the apparent height, and may indicate that a thinner pier is necessary to avoid a squat appearance.

Designers are often tempted in the name of economy to keep column diameters the same over several different bridges of differing heights, resulting in unattractive proportions in some of the structures. Some minimal cost savings may result, but the potential savings should be weighed against the loss of visual quality.

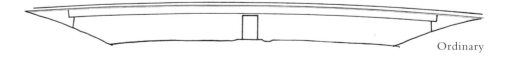

Ordinary

FIGURE 5-14 *Pier widths should relate to superstructure; proportions should take safety mounds into account.*

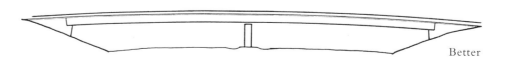

Better

Tall Piers

Tall piers are easier to design because both structure and aesthetics point in the same direction: consolidation of vertical members into one or two shafts. Engineer and laypersons alike can appreciate the economy of consolidating reactions into fewer members for the longer journey to the ground. Nature itself has offered a familiar model: the tree.

Use No More Than Two Columns per Pier at Each Pier Line

If more columns are necessary for unusually wide structures, they should be paired. Many columns together begin to look like a forest of columns. Pairing columns creates the illusion that there are fewer columns because each pair is seen as a single entity. See Figure 6-27.1 (page 219) for a good example.

Recognize and Accentuate the Verticality of Tall Piers

Tall piers provide an opportunity to create an appearance of slenderness. The key is to accentuate their verticality.

- Use simple, vertical shapes. If both horizontal and vertical members are present, emphasize the vertical members by making them larger or more continuous than the horizontal members. If the required width of pier at the top is small enough, a single vertical shaft will suffice (Figure 5-15). A single vertical shaft on a wide bridge will look—and be—too massive.

- Taper or flare the vertical members to be wider at the bottom. Because of wind, seismic, and other horizontal loads, the stresses are largest at the bottom; the pier should be as well. Piers will look too squat if the taper is too great. Tapers of 1:24 to 1:80 work well in most situations, with the lesser tapers applicable to taller piers (Figure 5-16).

- Integrate shaft and cap. The pier operates as one element structurally, so it ought to look like one element (Figure 5-17). The tree analogy, specifically the trunk/branch joint, can provide clues about the shaping of the shaft/pier cap intersection. The cap/branch should be thicker at the joint and thinner at the end. There is no structural reason to be restricted to straight lines. Curves more accurately reflect the flow of stresses, and they add visual life to the structure. If

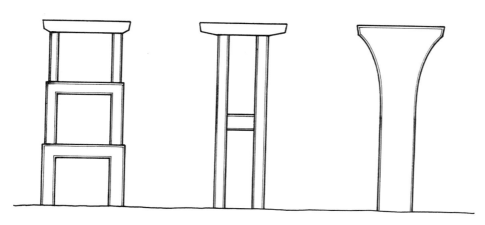

FIGURE 5-15 *The targets for tall piers: simplification and vertical emphasis.*

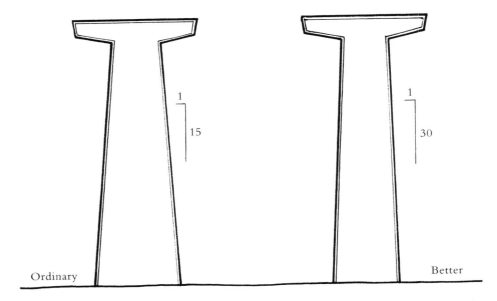

the same shape is used for the top section of all piers, the reuse of the form-work will make up for the additional cost of a curved shape.

There is a temptation to get carried away with this shaping process, and venture into the realm of nonstructural design. The shapes indicated by structural needs are exciting enough.

Families of Piers

Design a Group of Piers of Varying Heights as Variations of the Same Basic Shape

It is not uncommon for multi-span bridges to have piers of widely varying heights. Bridges over shipping channels or bridges over deep ravines are examples. One pier by itself may look fine, but many piers of differing heights seen at the same time will create visual cacophony. Simply changing from one type to another—for example, multicolumn bents for the short piers and hammerheads for the tall piers—

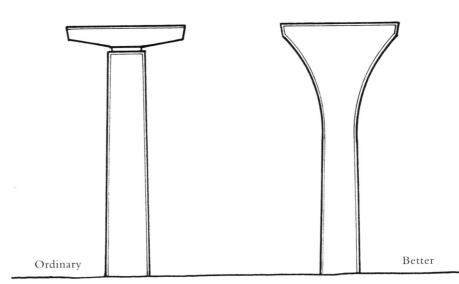

FIGURE 5-17 *The necessity of integrating pier cap and shaft.*

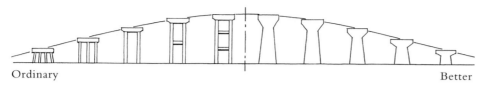

Ordinary   Better

FIGURE 5-18 *Families of piers varying by height.*

will always look like a camel of a compromise. The designer is challenged with creating a family of pier designs that look good both individually and as a group, whether the individual pier be tall or short. Usually it is better to pick a single type—say, a hammerhead or a two-column frame—and vary its proportions through the different piers in a logical and continuous way (Figure 5-18).

Design a Group of Piers of Varying Widths as Assemblies of the Same Basic Shape

An even more difficult problem is introduced when a bridge varies in width over multiple spans or branches, as when a ramp leaves a main line (Figure 5-19).

- With hammerhead piers, wider piers can be made by assembling a series of hammerheads placed tip-to-tip. As the bridge widens or branches, it is fairly simple matter to add another hammerhead to carry the additional structure. The side-by-side hammerheads can be connected structurally to improve their efficiency without changing the basic concept or its visual impact. Since they then become rigid frames, their shape continues to be consistent with structural logic.

- With a one-column-per-girder arrangement, more girders can be handled just by adding more columns. This works well with widely spaced girders, especially box girders. With closely spaced girders, it can degenerate into a forest of columns.

- Consider integral pier caps, which have major visual advantages in this situation (Figure 5-20.1). The bridge becomes a wide structural ribbon riding above a few simply shaped columns, with considerable flexibility in their placement on the ground (Figure 5-20.2). There are no pier caps to block oblique and longitudinal views through the structure.

Keep the Slopes and Curves of Adjoining Piers Consistent

This approach will result in piers that create pleasing compositions when seen together from the major viewpoints. When designing a series of piers that will be

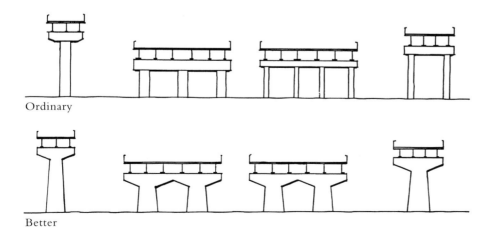

Ordinary

Better

FIGURE 5-19 *Families of piers varying by width.*

FIGURE 5-20.1 *These piers accommodate multiple widths with consistent shapes and details.*

FIGURE 5-20.2 *Integral pier caps provide flexibility to match ground-plane changes while maintaining a consistent appearance of the superstructure.*

seen at an oblique angle, so that they line up one behind the other, the shape of the visual voids between the piers should be considered, as well as the shape of the piers themselves. The Bong Bridge in Duluth, Minnesota, is a premier example of composing consistently (but not identically) shaped piers so that the group looks good from any angle (Figures 5-21 and 5-22).

FIGURE 5-21 *The piers of the Bong Bridge in Duluth, Minnesota, display a series of variations of the same basic shape to accommodate roadways of different height and width.*

FIGURE 5-22 *The shape chosen for the Bong Bridge piers creates interesting compositions from many angles; even the voids between the piers are consistently shaped.*

Nuances of Pier Shape

Keep Pier Designs Simple

Most piers are seen in groups by people traveling at high speeds who have little opportunity to understand and appreciate complexity, so keeping each pier simple is usually the best strategy.

- Use a minimum number of elements in the pier design. One shaft is better than two columns; two columns are better than three, and so on. For structures that stay low to the water, single shafts or walls will be best of all.

- Use simple patterns of large, easily seen elements.

Use a Wall or Fan-Shaped Pier under a Haunched Girder

Haunched girders resting on cantilevered pier caps can seem insufficiently supported. The curves of the haunch direct the eye toward the ground. The eye expects a solid support under the point of the haunch, continuing the line groundward (Figure 5-23). If the support is not there, as in a cantilevered pier cap end (Figure 5-24), the viewer will feel that the girder is hanging in midair.

As discussed in the "Superstructure Details" section in Chapter Four, the joint between pier and superstructure is crucial to the appearance of the bridge, as well as its performance. The joint should be designed to make clear its structural function. That means that a joint with a bearing will look noticeably different from a moment connection.

At Bearings Taper Visually Pier Tops and Chamfer Bearing Pads So That the Superstructure Stands Clear of the Substructure

There is no moment at a bearing, only axial compression and some shear, so the dimensions can be relatively small (Figure 5-25; also see Figure 5-9, page 148). If the superstructure appears to rest on a point, it will appear lighter and thinner than it really is. This technique will also accentuate the horizontal sweep of the superstructure. The effect is the strongest when an observer can see right through the structure between the girder and the substructure.

The top of the pier cap can be chamfered to assist in this effect. A chamfer also reflects the stress transition from the small base plate to the larger area of the pier.

FIGURE 5-23 *The haunch seems inadequately supported on this railroad bridge.*

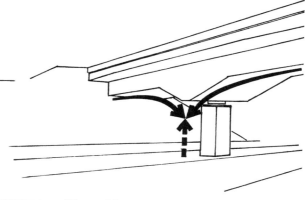

FIGURE 5-24 *The eye follows the lines of the haunch to the pier and expects to find something there to resist the forces implied.*

FIGURE 5-25 *The bearings allow a bit of sky to show through between the superstructure and substructure of this railroad bridge in Paris, emphasizing the continuity of the superstructure and making it seem light in weight.*

Finally, a chamfer or taper brings the pier to a clear visual termination, so it does not appear to be a short piece of a longer shape (cylinder, etc.), which happened to be sliced off by the superstructure.

For Monolithic Piers, Blend the Pier Shape Smoothly into the Superstructure

Exactly the opposite situation exists for piers that are monolithic with their superstructures. Here there is usually a substantial moment. in addition to compression and shear. The pier should blend smoothly into the superstructure to provide sufficient carrying capacity and to avoid points–of–stress concentration. If the same detail is used throughout the structure, the repetition will reduce the costs of the curved forms.

Create Consistency in Shape Among the Pier, Girder, and Parapet

In order to achieve unity in the structure, the shapes of the major elements should have some common characteristic (Figures 5-26.1 through 5-26.3).

Use piers formed with planes and straight edges with precast concrete I-girders, and rectilinear box girders, which are formed of planes. Use piers with curved surfaces with box girders having curved surfaces.

The I-shape of steel wide-flange and welded plate girders is less dominant visu-

FIGURE 5-26.1 THROUGH 5-26.3 *Sample techniques for making piers appear thinner; note consistency with parapet shapes.*

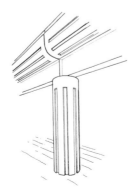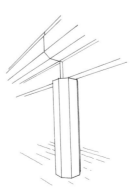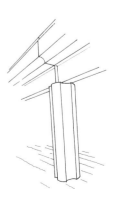

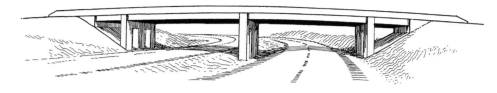

FIGURE 5-27 *Pilasters break up the horizontal line of a bridge and make it appear thicker.*

ally than other girder types. The choice of pier and parapet shape can be made based on other considerations. However, the pier and parapet should still share a commonality of shape.

Use Facets, Curves, Grooves, and/or Recessed Panels to Reduce Apparent Thickness of Piers

Where appropriate as part of the overall design, columns can be made to appear thinner by breaking them up into multiple surfaces with facets, curves, recessed panels, or vertical incisions (see Figures 5-20.1 and 5-20.2, page 154). The technique used should be consistent with the details used on of other parts of the structure.

Avoid Pilasters at Pier Caps

Pilasters or closure walls are sometimes used on pier caps to hide the bearings. They have the effect of interrupting the horizontal sweep of the bridge, breaking it up into segments, which thereby appear thicker than they otherwise would (Figure 5-27). Because the bearings are hidden, the viewer is left in doubt as to how the bridge is supported.

Design Pier Protection and Pier Together

When a pier is in a narrow median or next to a shoulder, protection for errant vehicles must be provided. The standard answer is a guardrail, which introduces still another element into the visual field, while changing the appearance and proportion of the pier (as well as introducing hazards of its own). Better approaches should be found.

- Use a tapered mound to deflect vehicles.
- Integrate the pier design with a Jersey barrier, as shown in Figure 5-2

FIGURE 5-28 *The piers of the Alberta, Canada, Deerfield Trail Interchange bridge are integrated into the median barriers. The opening between the slanted pier legs keeps the bridge open and transparent, and the parapets reflect the angles of the pier legs. This is a very successful illustration of the value of integrating the shapes of all the elements of a bridge.*

With both of these techniques, the thickness of the pier should be reduced to reflect the shorter apparent height of the pier and the consequent change in proportions that will result.

ABUTMENT AND WALL SHAPE

Abutments and adjoining retaining walls can be major contributors to the appearance of a structure, especially for structures of four or fewer spans. The shapes and details should be chosen to enhance the design intention and the features of the rest of the structure. Independent retaining walls and noise walls can also be major features in the highway scene. Their appearance deserves attention.

Abutments

The slope of the abutment face can strongly influence the appearance of bridges of up to four spans. The choice should depend on the visual intention of the design.

Slope the Face of the Abutment Outward to Emphasize the Continuity of the Space of the Undercrossing Roadway—and Its Adjoining Land Forms
If the abutment face is given sufficient outward slope (at least 1:2 vertical), the resulting batter will make the abutments appear more a part of the approach fills and will create a feeling of spaciousness under the bridge. This is a good approach when the intention is to deemphasize the importance of the over-crossing roadway and the bridge.

Slope the Face of the Abutment Inward to Create a Sense of Continuity between Abutment and Superstructure and to Accentuate the lines of the Crossing Roadway
If the abutment face is given sufficient inward slope (at least 1:2 vertical), the resulting cantilever will appear to be an extension of the superstructure and make the over crossing roadway and bridge appear more important.

Make the Face of the Abutment Vertical, to Create a Static, Formal Impression
If the abutment face is vertical, its lines are neutral and it adds little to the overall impression (Figures 5-29.1 through 5-29.3).

Consider Terracing Large Abutments with Planters
Adding terraces to a high abutment divides the wall vertically and makes its height less apparent. This approach is particularly effective in integrating the bridge with an adjoining formal landscaping. Such abutments can also give an ordinary highway overcrossing more visual presence, converting it into a gateway (Figures 5-30 and 5-31; see also, Figure 2-55, page 57).

End Parapet and Railings in Ways That Support the Design Intention
The parapet, railing, and pedestrian screens come to their ends at the abutment. How these elements end will have a major impact on appearance. There are three options:

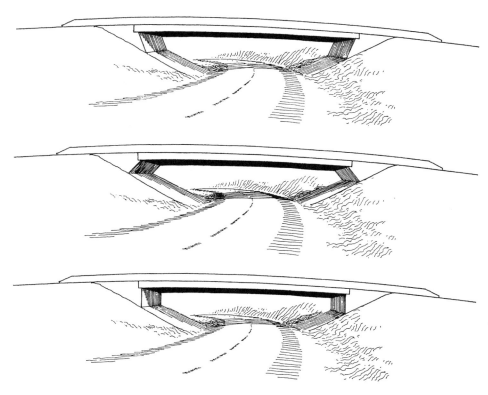

FIGURE 5-29.1 *Sloping the abutment face out to emphasizes the space of the under crossing roadway is an important factor in creating the aesthetic impression.*

FIGURE 5-29.2 *Sloping the abutment face inward emphasizes the lines of the overcrossing roadway and frames the opening.*

FIGURE 5-29.3 *A vertical abutment face presents a static, formal, visual image.*

- *Type 1.* Carrying the parapet profile and as much deck overhang as feasible across the abutment to its end will increase the apparent length of the bridge and make the superstructure appear thinner. This idea works best for abutments no higher than one-third of the height of the end span. For abutments higher than that, the high end wall dominates and offsets the effect. Essentially, the designer is trying to de-emphasize, by visual technique alone, two objects that visually comprise a major part of the structure (Figure 5-32).

- *Type 2.* Carrying the parapet profile down the end wall corner will make the abutment appear to frame the opening. This idea works best when there is a view beyond that is worth framing, with rigid frames where the girder and abutment face are structurally continuous, and in more urban situations where it can be coordinated with adjoining retaining walls (Figure 5-33).

- *Type 3.* Ending the parapet profile at the abutment end wall is the simplest option, but doing so will shorten the apparent length of the bridge and make

FIGURE 5-30 *Terraces can reduce the apparent height of an abutment and integrate it with a landscaping plan; Polaris Interchange, I-70, Columbus, Ohio.*

FIGURE 5-31 *Avery Road over U.S. 33, Dublin, Ohio. These terraces integrate the bridge with an interchange landscaping plan aimed at creating a gateway for Dublin.*

FIGURE 5-32 *Example of type 1.*

FIGURE 5-33 *Example of type 2.*

FIGURE 5-34 *Example of type 3.*

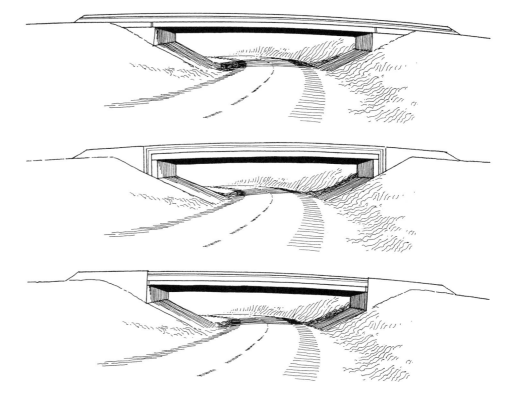

the superstructure appear deeper. It also appears to leave the girder unsupported (Figure 5-34).

The choice between alternatives should be based on the designer's design intention for the structure. Finally, as the height decreases toward pedestal abutments, the differences between them disappear.

If Bearings Occur at the Abutment, Leave Them Exposed

Bearings at abutments can be exposed or hidden behind a "cheek" wall. This decision often provokes passionate debate. Many feel that bearings should always be hidden because doing so simplifies the overall appearance of the structure and leaves the eye free to concentrate on the pier or other features. Disagreeing, an outspoken minority feel bearings should be exposed because the increment of additional length will make the bridge seem thinner and demonstrate how the superstructure is supported. The author's opinion is that the question should be answered by the nature of the girder abutment joint: When the girder is acting integrally with the abutment, as in jointless bridges and abutment-restrained girders, the joint should be hidden. Otherwise, the bearing should be exposed (Figure 5-35.1).

Make Beam Seat Width of at Least One-Half the Girder Depth

Narrow beam seats will seem visually too small to support the weight of the girders.

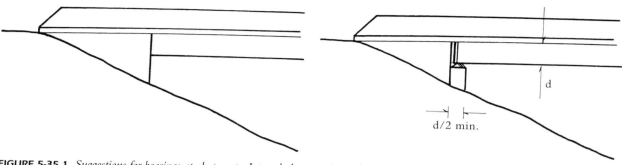

FIGURE 5-35.1 *Suggestions for bearings at abutments. Integral abutment is on the left.*

Recess the Beam Seat under the Girder to the Same Plane as the Fascia Girder

If the area of abutment under the girder is recessed to the same plane as the girder, the apparent height of the abutment will be reduced.

Design Abutments That Adjoin Retaining Walls as a Continuation of the Retaining Walls

Structures often occur in proximity to retaining walls. The wall should blend into the abutment without abrupt changes in pattern or configuration (see also Figure 5–36 and Figure 4–38, page 122).

Retaining Walls and Abutment Wing Walls

The shape of retaining walls and noise walls is determined, to a large degree, by their placement in the highway environment, which determines their height at each point and, therefore, their overall shape as seen by drivers and other viewers. Full discussion of the factors involved was provided in Chapter Four in the "Abutment and Wall Placement" section. Following the advice in that section should result in a continuously curved plan and top profile, without jogs, offsets, or steps. The following guidelines focus on the faces and details of walls.

Batter the Faces of Retaining Walls and Abutment Wing Walls

The tops of vertical walls, especially high ones, appear to lean outward, as if they were about to topple over. A slight batter eliminates this effect at the same time it indulges our intuitive feeling that walls need to be thicker at the bottom to resist the earth behind them. The backs of walls are often battered, so why not turn the wall around and let the batter show, as in Figure 5-30?

How to trim the wall at the top is a key decision:

If Natural Surfaces Will Be Visible above the Wall, End It without a Coping or Cap

The wall will seem more a part of the landscape if there is no cap to create a line between the two.

If Other Structures, Roadways, or Buildings Will Be Visible, the Wall Should Be Topped with a Cap or Coping

Parapets, railings, and fencing installed on retaining walls should follow the same guidelines as for the same elements on bridges. If there are bridges near or adjacent to the wall, the parapet, railing, and fencing details should match (see Figure 5-36).

Use a Cast-in-Place Cap with Precast Retaining Wall Units to Achieve a Smoothly Curved Top Profile and to End the Geometric Pattern of the Units

Retaining walls made of proprietary, repetitive, and precast elements have characteristic geometric patterns resulting from the stacking of the individual elements. If the wall height were an even multiple of one of these units, then the top edge would be smooth; but that never happens. Instead, the top course is typically made up of fractional units or has large steps when the top edge jumps from one course to the next. As in all walls, the most important thing is to provide a smoothly curved alignment and top profile, which can be done with a cast-in-place cap.

Walls made up of precast crib units, large precast planting boxes, and gabion walls offer another alternative. A smoothly curved alignment (within the restrictions of the unit size) is again a requirement. However, the dimension of the units usually prevents a smoothly curved top profile. Conversely, the open nature of the construction offers possibilities for plant growth that can convert one of these walls into a kind of hanging garden. In fact, they should only be used where this kind of planting can be achieved. Then the top can be stepped back in a logical pattern related to the topography, and each step can be made a platform for further planting. (see Figure 5-52, page 180).

Visually Coordinate the Components of Soldier Pile Retaining Walls and Panelized Noise Walls

Both retaining walls and noise walls are often made up of vertical piles or posts with concrete panels between them. The appearance of these walls can vary tremendously depending on how much care is taken in getting the individual elements to look consistent and fit together. The walls look worst when the panels are made up of individual horizontal elements, producing a log-cabin look to the wall. There is too much visual competition between the horizontal lines of the panels and the vertical lines of the posts. Horizontal lines on the infill panels should be minimized by using large panels and/or by using vertical grooves or some other pattern on the panels to visually suppress the horizontal lines (see Figure 5-53.1 and 5-53.2, page 181).

Visually De-emphasize the Piles/Posts of Panelized Walls

The prominence of the piles/post every 12 or so feet can also be a problem, creating a monotonous repetition of closely spaced verticals that stretches out for thousands of feet. The effect can be reduced using one of these methods:

*   Minimize pile/post width.

*   Mount one adjoining panel to the front and the other to the back of the post in zig-zag fashion, so that only the edge of one panel is seen.

*   Give the pile/post the same texture and color as the panels.

The installation of tieback anchorages on soldier piles often creates an irregular pattern on the face of the wall. Ideally, these should be controlled so that they all appear in rows that are horizontal or parallel to the vertical alignment. If that is not possible, they should be encased within cast-in-place pilasters, the tops of which are aligned in an even row.

Eliminate or Visually Minimize Steps in the Top Profile of Panelized Walls

The first goal should be to eliminate steps. There is no functional or cost reason why panels cannot be cast at an angle, so that the top forms a smooth extension of the panels on either side. However, if that is not possible, a combination of techniques can create the illusion of a continuous smooth top profile. First, make each step a consistent height of only 6 to 12 inches, then use a cap feature, such as a smooth band on a roughly textured panel, with a somewhat larger dimension.

Noise Walls on Retaining Walls

Noise walls are sometimes placed on the tops of retaining walls. These are often designed as two separate elements, resulting in an inconsistent and even conflicting appearance. Perhaps this happens because retaining walls are designed by the structure division, and noise walls are designed by the environmental mitigation division. The point is, travelers and adjoining residents see them together, so the two divisions should design them together. The problem can be complicated when the noise wall is on the retaining wall for just part of it length, and on an independent alignment elsewhere.

Try to Separate Noise Walls from Retaining Walls

Because noise walls and retaining walls have such different functions, the techniques of construction can differ significantly. This makes it very difficult to make them look integrated. So, the first choice should be to avoid placing a noise wall on top of a retaining wall. It is well worth the effort to look for an additional 6 feet or so to set the noise wall behind the retaining wall and create a planted buffer between the two. Even with such a minimal division, the two will be perceived as separate elements and can be designed individually. That said, some coordination of materials and colors would still be desirable.

Design Noise Walls on Retaining Walls Together

When the two walls must be on top of one another, more complete coordination is necessary.

- Line up and similarly size any vertical elements (piles/posts/pilasters).

- Use the same material for both, with the same color and texture.

- Align the tops of the two walls in parallel.

- If the retaining wall is battered, batter the noise wall.

The retaining wall/noise wall combination in Figure 5-50 (page 180) is an excellent example of how to combine the two.

Retaining and noise walls offer large surfaces that need to be visually organized, otherwise the surface will be the result of the vagaries of the construction process and good appearance will be unlikely and entirely coincidental. An infinite number of possibilities exist, so it is difficult to generalize. For further discussion of wall surface treatments, see the "Details, Surfaces, Textures, and Ornamentation" section, later in this chapter.

COLOR

While the strongest determinants of the visual impression of a bridge are the shapes of its major elements, the surfaces of those shapes can, through color or texture, alter our perceptions of them. Because of the size of bridges and the distances at which they are usually first seen, color is a very important influence on how well the shape is perceived. A bridge that is very similar in color to its background will not be seen as clearly as one whose color contrasts with its background (Figure 5-37.1 and 5-37.2). For all of these reasons, color can be a very effective way to enhance and enrich a structural form.

Pattern and texture are other characteristics of surfaces that affect people's impression of a structure. Traditional materials such as brick and stone have their own characteristic pattern. However, when considering a use of pattern and/or texture, it's important to understand the color implications first. Color will influence the impression *before* pattern and texture. At the distances at which color is first seen, pattern and texture will have an effect, but only as modifiers of color. For example, when seen at a distance, concrete with a surface formed by narrow grooves will appear as a darker shade of the natural concrete color. The impression of a darker shade is produced by the micro-shadows created by the grooves; the grooves themselves will not be identifiable. For this reason, this section focuses first on an exploration of color; the next section moves on to patterns, textures, and ornament.

The Role of Color

Color is an intrinsic part of people's perception of an object. Color will influence their reaction to that object, and that influence will occur whether or not the designer seeks to control the color.

FIGURE 5-37.1 *This image and the following one show the same bridge painted in two different colors, and illustrates how the color of a structure and its background affect our perception. In this image, the orange bridge contrasts strongly with the winter-time grays and browns of its background. (See color insert.)*

FIGURE 5-37.2 *In these images, the green bridge blends into its summer-time background. (See color insert.)*

Bridges are part of a larger visual scene, all parts of which have a color. The other colors of the scene are almost always beyond the control of the engineer. The bridge itself will have colors, even if they are simply the natural colors of the materials selected for structural reasons. Those colors will become an integral part of the scene to which the bridge is added.

The impression created by the bridge and the emotions evoked by it will be influenced by all of these colors as they are seen together. As we shall see, the impact a color makes—indeed, the hue and value we perceive the color to be—is affected by all of the other colors seen around it. It is important for engineers to understand these relationships in order to correctly anticipate the effect of their decisions about color.

Color Basics

Using color requires an understanding of light, how colors work together, and how people see color. The use of color then becomes a result of knowledge and awareness rather than intuition alone. There's a bonus from the effort of learning more

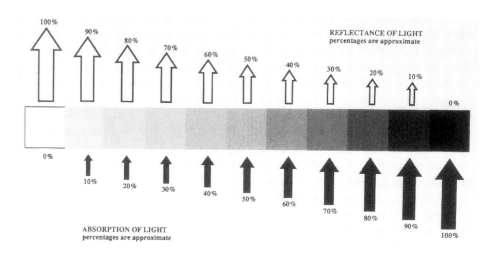

FIGURE 5-38 *The value spectrum.*

FIGURE 5-39 *Contrast in value is the dominant factor in making objects distinguishable from their backgrounds.*

about color: As knowledge increases, intuition becomes more effective. Let's start with some definitions.[1]

Value is the overall lightness or darkness of a surface. It is determined by the amount of absorption of all of the light striking the surface. It is usually measured by comparison to tones of gray, defined by the varying percentages of light they absorb. Values can be judged by squinting at the view. In the limited light, our eyes can no longer recognize hue, but they can recognize value (Figure 5-38). Indeed, we depend primarily on contrast in value, not color, to distinguish objects from their background (Figure 5-39).

Hue (red, yellow, blue) is determined by which wavelengths of light are absorbed/reflected by a material. Hue relationships can most easily be understood by curving the visible spectrum into a circle, with red placed next to violet, creating a color wheel. The most common version of this approach names red, yellow, and blue as *primary hues,* hues that cannot be mixed from other hues. The primary hues lie about 120 degrees apart on the color circle (Figure 5-40). The *secondary hues* are mixtures of the primary hues on either side of themselves: orange equals red plus yellow; green equals yellow plus blue; violet equals blue plus red. A large number of *intermediary hues* between the primary and secondary hues can be mixed, depending on the proportions of primary hues used. Hues at 180 degrees from each other on the color wheel are called *complementary hues.* Red/green, yellow/violet, and blue/orange are complementary hues. Each intermediate hue also has its own complement. Placing complementary hues adjacent to each other creates the strongest reactions and the greatest contrasts of any color combinations.

Intensity is the relative degree of purity of a color. Other terms also used for this color characteristic: purity, saturation, strength, chroma, and brightness/dullness. *Tints* of a hue may be made by mixing it with white. *Shades* of a hue may be made with mixing it with black. Tints and shades reduce intensity. Hues can also be reduced in intensity by mixing them with their complement, which will eventually produce a tertiary color.

Tertiary colors are chromatic neutrals. They may contain any and all hue ingredients but with no single hue easily identifiable. They cannot be identified as a hue, but neither are they a simple mixture of black and white. "Brown" is often used to describe many of the colors in this "not black, gray, or identifiable-color" family. However, in most people's minds, brown has a red or orange content, which leaves us without words for the rich variety of tertiary colors that are less red and more green or blue in tone. Many of the colors of nature, such as soils, rock faces, and tree bark, are low-intensity shades of tertiary colors, often quite different from "brown." The harmony of these colors with other hues will depend in large part on whether there is an identifiable base hue. If there is, the rules of harmony will be the same as for that base hue. Thus, it is important to be able to pick out any identifiable base hue in tertiary colors in a natural background in order to judge the relation of the background to proposed bridge colors. If there is no identifiable base hue, then the color will match with anything.

Three-dimensional color wheels are available that relate all colors according to hue and intensity.[2] The wheels place low-intensity colors like tan and beige into a clear relationship with the pure colors. They are important tools in learning to judge the hue components of natural colors.

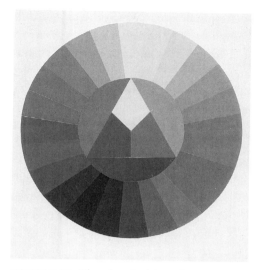

FIGURE 5-40 *The color wheel (see color insert).*

FIGURE 5-41 *The center panels show tertiary colors in which no hue can be discerned (see color insert).*

Temperature is a color characteristic that produces psychological and, to some extent, physical sensations of warmth or coolness. Red, orange, and yellow are considered the *warm hues*. Green, blue, and violet are considered the *cool hues*. At the boundaries between green and yellow and between violet and red are intermediate hues, which seem either warm or cool, depending on their backgrounds.

Characteristics of Backgrounds

As discussed in Chapter Two, the daylight in each part of the country has its own characteristics, some obvious, some subtle. This is the starting point for decisions about color. Locations with stronger, brighter light will create strong shadows and allow darker colors to be perceived. Lighter colors may seem washed out. Weaker light results in fuzzier shadows with less contrast (Figures 1-8 [page 4] and 5-42). Colors must be clearer and lighter for the hue to be perceived.

FIGURE 5-42.1 *Ridge Road over I-70 near Frederick, Maryland, photographed in the late afternoon in October, looking from the west (see color insert).*

FIGURE 5-42.2 *Ridge Road over I-70 near Frederick, Maryland, photographed in the late afternoon in October looking from the east (see color insert).*

The colors of the site constitute the background against which the bridge will be seen. The value(s) and hue(s) of the background will be present over a much larger area of the visual field than the area of the structure. It is important to understand what they are. Different parts of the country have different characteristic backgrounds, generally dominated by the vegetation or lack thereof. In the Northeast, Mid-Atlantic, Southeast, and Midwest, the background will be multiple shades of green in spring and summer, and briefly reds, oranges, and yellows in the fall. Wintertime backgrounds are dominated by the colors of intertwined tree trunks and branches and dormant ground vegetation. Many people think of the resulting color as "brown," but in fact it is a rich mixture of tertiary colors that appear closer in total effect to a warm gray. Across the Southwest and West, the background will be the sage green, brown, and gray of desert vegetation and topography. The color chosen for the Big I interchange shown in Figure 6-59 (page 248) works well because it picks up the rose beige of nearby mountain faces. In the Northwest, the dominant color is usually the dark green of the coniferous forest. In California, it is a few months of spring green followed by months of golden brown.

It's important to see what colors really *are,* as opposed to what you think they are *supposed* to be. Many people tend to feel colors should follow preconceived rules. Trees are green, which to many people, is the green that came in their elementary school crayon box. But consider that a copper beech tree is not green, it is purple. In fact, hundreds of shades of green are necessary to depict all of the trees native to the United States.

The preceding general statements may or may not apply to a specific site. Judgments about a specific bridge can be made based only on the conditions at its site. Begin with the important viewpoints identified earlier. From those locations, look at the values and hues of the background against which the bridge will be placed. Take color photos. Don't forget seasonal changes; in most of the country, there are well-defined seasons. If there will be snow on the ground for five months, get some wintertime shots, too. Try to get the background colors as accurate as you can, and use them in making judgments about color.

Some caution needs to be exercised regarding color photos. Many color films do not show every color to its true intensity and temperature. The same can be true of digital cameras or the computer monitor/printer on which the images are viewed/printed. Shoot some experimental images with the type of film or equipment that you expect to use. Then compare the results to the actual scene viewed in similar light conditions. The comparison will give an indication whether any "mental corrections" will be necessary.

Understand the position of the bridge in the visual field from the various viewpoints. How much of the field does it occupy? The larger the proportion of the visual field that it occupies, the less influence the background color will have on the final result. Also consider the orientation of the bridge. Colors will appear brighter on south-facing surfaces and, for half the day, on east- and west-facing surfaces. They will also fade faster on these surfaces. Figures 5-42.1 and 5-42.2 show the different impressions created by the same structure at different times of the day.

If the bridge or its immediate environs are lit, consider nighttime conditions. These can include the exaggeration of shadows and textures, depending on the location of the light source or sources, and changes in apparent color caused by the color effect of different types of lighting. For example, mercury vapor and metal halide lighting will bring out blues and greens, while reds and oranges become dulled and grayish; high-pressure sodium lighting will bring out reds and oranges, while blues and greens will become dulled and grayish.

Because signs have characteristic, strong colors, the presence of a sign on a bridge must influence the color selection. This is another good reason to keep signs off bridges.

Choosing Colors

In choosing colors for a bridge, the designer must focus on picking the color(s) of the bridge that will best combine with the background to evoke his or her design intent.

Determine the Color Effect Desired to Meet the Design Intent

There are a number of effects that can be achieved with color:

- To make a bridge blend in or disappear against the background, either to avoid drawing attention from a more important object, or to prevent the visual cacophony of too many different colors in the visual field. Many highway bridges are not intended to be the focus of a given area. As background structures, their colors are better kept simple and subdued.

- To make the bridge contrast with its background, or to bring out the bridge as a feature, a source of excitement, pride, or orientation. Many highway environments are spread out and monotonous. A bridge can be a feature that adds a moment of pleasure to a journey.

- To provide variety and interest, perhaps as part of a larger theme in, say, an urban commercial area with an overall urban design plan.

- To provide a symbol of an institution, town, or historic district, by using colors associated with that entity.

- To emphasize the structural shape itself by giving it a color that contrasts strongly with its background. The brilliant red-orange of the Golden Gate Bridge towering against the blue sky is a familiar example (see Figure 1-2, page 2).

- To display and underline the workings of the structure by using different colors for different parts of a structure. The Victorians carried this to an effective extreme in their iron structures, right down to a different color for the bolt heads.

- To create a desired emotional effect. Different color combinations (bridge plus background) can evoke different emotional effects. Strong contrasts, especially in the values and intensities used, will tend to be more stimulating and exciting; weak contrasts will impart a more soothing, quiet effect.

Color can be an effective method to reach these goals in a way that does not compromise the structural form or add undue cost. Not all of these goals are mutually exclusive; several can be effectively applied in combination.

Once a design intention is determined, the designer can move on to value and hue considerations. Since value comparisons dominate our initial impression, it is critical that the desired value combination be determined first.

Settle on a Desired Value Scheme

The balance of light against dark is the first decision to make. Most natural backgrounds tend toward the middle range, so medium values will help the structure blend in. Much lighter or darker values will be necessary to make the structure stand out. Much value contrast is the result of shadows and shades cast on one part of the bridge by another part, and these need to be anticipated.

Evaluate the Effect of Background Colors on Potential Hue Schemes

If the desire is to blend the structure into the scene, colors should be used that are similar to the background colors. The choice of colors should usually be based on the longest period of similar colors, usually spring-summer. Decide which part of the background the structure should blend into. The structures in Figure 5-44, photographed during painting, interacts with their background in two different ways. The closer, silver-gray structure takes on the color of the sky and seems to be pulled up into the sky; the rear structure, because of its value and hue, seems a part of the land and seems to visually connect the two hillsides.

Determine Whether the Structural Materials Will Be Coated or Colored

Concrete and A588 (weathering) steel have characteristic colors in their natural states. With respect to each of the surfaces of the structure, a decision must be made whether to leave the structural material as-is, to add some additional color, texture, or pattern, or cover the surface with another material. Keep in mind that a decision to leave a structural material in its natural state *is* a decision about color, whatever other criteria or considerations may have motivated it.

Two major points to consider:

- The application of color or texture is not *necessary* for the creation of a good-looking bridge; appropriately shaped structural materials in their natural state

FIGURE 5-43 *Weak-value contrast, usually a result of haze or cloudy conditions; even the dark-gray shadow on the girder contains discernible detail (see color insert). Compare this with the strong-value contrasts in Figure 1-8; in the strong, high-altitude Colorado sunshine, even the light-colored abutment concrete is seen as black within the shadowed area (see color insert).*

FIGURE 5-44 *Photographed halfway through a repainting, the Maury River Bridge in Virginia demonstrates the interaction of color and background: The aluminum-colored span takes on the sky color and seems part of the sky, while the brown span emulates the colors of the topography and clearly ties together the two hillsides (see color insert).*

can do that on their own. Color and texture are bonuses, sources of enrichment and interest that can enhance a good structural design.

- Conversely, the application of color or texture is not *sufficient* to make up for an ugly structural form. Many other bridges are painted red-orange, but it is the Golden Gate Bridge that has captured our imagination. Color can be used to influence our perception of shape, and thereby improve the appearance of an ordinary bridge, but it can never entirely compensate for poor decisions made about the basic shapes. Perhaps the best that can be done with color on an ugly bridge is to paint it the same color as its background, in the hope that it will be harder to see!

Most bridge-building agencies have a settled policy regarding the coating or coloring of structural materials, based on cost and maintenance considerations. Because both concrete and A588 steel are reasonably durable in most situations, frequently the policy is to *not* coat either one.

A serious problem can occur when a weathering steel superstructure is combined with uncoated concrete substructures. Runoff from the steel stains everything below it for years until the steel surface stabilizes. By then, the concrete substructures will be streaked almost as dark as the steel. One technique sometimes used to address this problem is to paint the section of the girder above the pier a color intended to match the natural steel color. However, the match never works beyond the first few weeks; the paint fades while the weathering steel darkens. Thus, the contrast is soon apparent and visually breaks up the girder into sections, making it appear shorter, deeper, and discontinuous. Better solutions are available. Either the runoff must be caught and directed away from concrete surfaces or the concrete must be coated.

These considerations may effectively establish the colors of the major structural elements without any color input from the designer. However, if the natural colors will not match the design intention, the designer must decide whether the bridge can be made to match the design intention through the application of color to selected details. If that is not possible, the designer should appeal for an exception to the no-coating policy.

If the agency is open to, but does not require, coating or coloring of the structural materials, a decision must still be made on whether that is desirable on aesthetic grounds. Uncoated, uncolored concrete and A588 steel have characteristic colors that lend themselves to certain natural backgrounds. Leaving them alone may be the best choice. Only if the agency *requires* coating can the following step be skipped.

Evaluate the Effect of the Natural Colors of the Structural Materials

What is the effect of the natural color of the structural materials against the relevant backgrounds? What are the visual/emotional effects that are likely to be evoked? Do those match the intention of the designer?

- A588 steel's black/brown color will dominate all other colors in the structure. The material starts as a medium brown, but over a 10-year period reaches a permanent black/brown. There are even hints of purple in the final hue. The exact final color depends on local atmospheric conditions. This aggressive color strongly affects all other colors used in the structure. The dominance is a

FIGURE 5-45 *This red railing adds interest to, and unifies, the mottled concrete of this structure (see color insert).*

result of both the strong hue and the dark value. While there are several possibilities for dealing with the problem, it is complicated by the fact that the steel color is a moving target. It starts out a rich reddish-brown but gradually loses hue as it approaches black. The material can fit in well in wooded, natural, and industrial areas. It is more of a problem in urban areas, especially if there will be pedestrian use around and under the structure. It is unlikely to be compatible with nearby buildings, and the dark color absorbs so much light that it is difficult to effectively light the underside.

Small areas or details of intense color on some other part of the structure can help offset these difficulties. The color could be applied as a "racing stripe" on the parapet, on specific areas of the piers or abutments, or on a railing or pedestrian screen (Figure 5-45). The effect will be to lessen the intensity of the steel hue and value. Red, violet, or blue details would tend to make the structure stand out against most backgrounds, with the possible exception of a blue seen against the sky or water. If the intention is to make the structure blend in, neutrals of similar hue to the background and of a medium to dark value could be used in the same way.

• Concrete's light gray-tan color will visually contrast with most backgrounds, and the structure will stand out. In particularly sensitive natural areas, concrete has been colored to blend in with the natural background. In urban areas, most people consider concrete's natural color acceptable, if bland. The reflective ability of the lighter color will make the underside more pleasant for nearby pedestrians. One problem is the lightly mottled appearance caused by slight differences in the concrete batches and/or imperfections in the forms. Also, the color is not memorable, and does little to add interest or life to the structure. Here again, insets or details of stronger, brighter colors can help add life to the structure by attracting attention while the concrete becomes the frame (Figure 5-46).

If, as a result of this analysis, a decision is made that requires coating the structural materials, the designer can go on to consider the specific hues and values to be applied to the structural materials to accomplish the design intention.

Consider the Hue/Value Combinations Available with Likely Coating/Coloring Systems

There are significant differences in the application of color to concrete as opposed to steel. With steel bridges, a wide variety of paint colors are available and colors can be picked to fit a color intention. Quality control is relatively easy to achieve, and the need for periodic repainting for maintenance reasons means that the color will be periodically renewed (though perhaps not in an unusual color or pattern). None of these factors apply to concrete.

There are three possible approaches to coloring concrete:

- Integrally colored concrete is the most durable, but the colors available tend to be limited, and generally are earth tones: siennas, golds, and umbers. Uniformity requires careful quality control, and can be difficult to achieve. However, some random irregularity can be an advantage when the goal is to blend the structure into a natural background.

- Staining concrete is another possibility, though the range of pigments is also limited, and it produces a mottled effect. This technique is really only useful when the mottled effect is desirable.

- External coatings are the most promising approach to coloring concrete and can be quite durable if correctly applied. They have an additional advantage of not requiring as much quality in the concrete finish as it comes from the form, since the coating will cover minor blemishes.

A basic problem with all concrete color is that the basic pigments are not durable. Since it is not necessary, for maintenance reasons, to recoat concrete, the

FIGURE 5-46 *Using a colored inset to add a strip of bright color to a concrete structure (see color insert).*

material will continue to exist in its faded condition. The earth tones and blues tend to last the longest; reds go the quickest. Color on the south side of the bridge will fade noticeably more quickly than color on the north side. Lighter colors will lose less to fading, and chipping or flaking will not be as noticeable. The likely degree of maintenance is also a major consideration. Colored or textured colored concrete in locations subject to vehicular impact will be a particular problem, as it is almost impossible to repair it to match the original. If the coating is a light color similar to concrete, any future damage will be less noticeable.

Aluminum is often used for pedestrian and bicycle railings. The standard mill finish weathers to a light gray oxide. However, aluminum can be painted in any color. It can also be anodized, which is more durable but which has a more restricted range of colors. Galvanized steel also weathers to a light gray oxide. Painting galvanized steel is possible, but requires strict quality control.

Pick Values and Hues in Relation to the Backgrounds the Bridge Will Be Seen Against When Viewed from the Important Viewpoints

- If the goal is to have the bridge blend in, the bridge value should be similar to the background value; bridge hues should be similar or close color-wheel neighbors of the dominant background colors. Take care when picking greens to blend into a natural background. There are many shades of green within the foliage of a single tree, and they change constantly with light and wind conditions. Picking a single green that will look good against a group of different types of trees is difficult. Good color photos and visits to the site under varied light conditions can help.

- If the goal is to have the structure stand out against its background, the most effective technique will be a strong contrast of hue and/or value. The use of colors that are complementary to the background colors will have the most striking effect. The bridge in Figure 5-42.1 (page 167) stands out because its bright red-orange is a near complement to the green hillside. Using one or two colors that are color-wheel neighbors of the complementary color is also effective.

The two versions of the same bridge shown in Figure 5-37 (page 165) illustrate both strategies.

Visualizations of the options must be made using the actual hues and values; words won't do. There are too many varieties of green for "use green" to be of much help. Begin to experiment with rough studies made with color photographs of the site as the background. Use colored markers or pencils.[3] Don't worry about details. It is the large areas of color that will create the effect. If CAD is available, a three-dimensional model of the structure can be placed against a scanned photo, and colors manipulated with image software.

When Picking Colors, Remember the Public Reacts More Positively to Brighter, Clearer Colors (Blue, Red, Orange, Green) and Less Positively to Silver, Gray, Brown, and Black

William Zuk surveyed the reactions of thousands of drivers to the same bridge painted different colors; a clear preference emerged for the stronger colors.[4] And Tom Porter and Bryon Mikellides, in their book, *Color for Architecture,* showed that the people for whom we presumably design have long desired more and better color in their built environments.[5] In choosing the color for the bridge shown in Figure 4-7 (page 102), the Pennsylvania Turnpike Commission picked a bright blue that contrasts with the pervasive green of its surroundings, ties the bridge to the sky, and makes this high abutment bridge seem more open. It also creates a memorable visual event to mark the long journey through the Appalachians.

Look for Opportunities to Incorporate Bright Color Accents

A painted "racing stripe" in bright blue was a very effective addition to the parapets of the Big I interchange (See the case study in Chapter Six). Such details are particularly effective where the structural materials will be left in their natural state. Where strong, permanent colors are desired as part of a surface design in concrete, a better approach than paint is to inlay other materials with permanent, characteristic colors. Examples are terracotta, tile, and glazed ceramic tile. These are available in a wide variety of colors, and have centuries of successful exterior use in architecture. For example, one might conceive of working a decorative pattern built around the colors of a university into a parapet of a nearby bridge.

If a color covers only a small portion of any one element of the bridge, and it is clearly incorporated into a decorative pattern independent of the structural form of the bridge, significantly more latitude is available in the choice of colors. This can be a tricky business, however. Large-scale full-color drawings are necessary to judge the effect, and professional advice may be necessary. For example, an artist might be asked to design an inset of colored tile for a parapet or retaining wall.

Consider the Colors of Other Nearby Structures

When groups of closely spaced bridges are involved, they will be seen one after another in quick succession. The colors for each should be picked in relation to the others. This does not mean that they all have to be the same, but that they all have some discernible relationship. For example, one might conceive of a series of bridges on an urban freeway being painted in a series of colors that gradually shift through the color spectrum of the various shades of red; then follow that group with another group that shifts through the various shades of blue.

Consider the Presence of a Sign When Making Color Selections

The green color of the national standard guide sign is not found in nature. The pure primary and secondary hues (other than green) harmonize best with it. Combining it with other hues just reinforces the artificiality of the sign color. This problem is another good reason to avoid signs on bridges.

When Choosing Colors for Pedestrian Screens, Sign Structures, and Light Poles, Consider the Background and the Design Intention

- To make these elements less obvious against the sky, use a light-gray/aluminum color. This color tends to pick up the sky color, so these elements will be harder to discern and will not distract from the main structural features.

- To make pedestrian screens, sign structures, and light poles less obvious against a hillside, use dark neutrals of a hue similar to the hillside, or black. Vegetated hillsides have a lot of shadow in the spaces between the leaves, and black elements seem to fade into them. Paint the rear of a bridge-mounted sign and its mounting structure a color that will make it blend into the background it is seen against. The back of the sign should be made to disappear as much as possible.

Be Aware of the Nuances of Color Selection

When it comes to the nuances of color, be aware of the following:

- Light colors result in stronger shadows, making any design that depends on contrasting shadows more effective. A dark-colored surface is already absorbing most of the incident light, so the effect of a shadow is less pronounced.

- Light-colored surfaces will reflect colored light onto nearby surfaces. This will be especially noticeable on surfaces that are in shadow and not receiving much direct light. For example, the soffit of a deck next to a bright red girder web will have a red tinge. This can be a bonus for structures where pedestrians use the space below. Conversely, dark-colored surfaces will reflect very little light onto nearby surfaces, making them seem dark as well.

- Avoid complementary colors on the same structure. While complementary colors can work well to establish a contrast with the background, it is risky to use them in the same structure. They will disturb the unity of the structure. Better to use similar colors within the same or nearby structures.

Select Retaining Wall Color as a Function of the Size of the Wall in the Visual Field and Color of Nearby Structures

Small walls seen primarily against natural features should probably be in colors that blend with nearby natural colors. Walls in a primarily man-made environment or that adjoin other structures should borrow colors from those structures. Walls are truly background structures, and rarely deserve the contrast in hue or value that would draw attention to them.

Select Noise Wall Colors That Will Blend the Wall into the Natural Background

Noise walls are usually seen against a background of planting or topography. Colors should be compatible with the seasonal variation of such backgrounds.

Use a Formal Color-Matching System to Specify the Colors Chosen

After the decisions have been made and the colors chosen, one important step remains: ensuring that the actual colors applied match the colors picked. There are important differences between the way colors are created for colored pencils, art markers, photographs, computer screens, color printers, and paint manufacturers. One medium's "green" is not the same as another's. One way to overcome these differences is to use a color-matching system, such as the Pantone system[6] or the U.S. government's Federal Standard 595B colors. These systems match colors to uniform color samples that are identified by number.

The best approach is to require the preparation on site of a sample panel, using the same substrate and field application techniques that will occur in the final structure. The sample can then be compared to the colors selected, and adjustments made.

Getting Advice

By this point, many engineers may be feeling overwhelmed by the complexity of this subject. For particularly complicated situations it is appropriate to call upon other visual professionals or specialists for advice as to color selection. Even trained visual professionals in fields such as architecture and interior design rely on color specialists from time to time. Engineers can, too.

These decisions can be complicated by the desire of community groups to participate in the process. Involving the community requires first narrowing the range of possibilities to two or three alternative color schemes. Presenting the public with the entire rainbow of possibilities will paralyze the endeavor. Each scheme should assign appropriate and consistent colors to all parts of the project. The community can then pick one of the schemes with the assurance that attractive and mutually consistent colors will result. The State of New Mexico developed two complete color schemes for the Big I Interchange, then presented them to the public via an Internet survey to make the final selection.

SURFACE PATTERNS, TEXTURE, ORNAMENTATION, AND DETAILS

Once color is determined, pattern, surfacing materials, ornamentation, and details can be studied. The two major goals of pattern/texture/ornament in a bridge should be:

- To differentiate the various parts of the structure's lines and thereby clarify them.

- To create a desired emotional response in the viewer.

Pattern and texture are difficult to perceive by people traveling at highway speeds. Only large and distinct patterns will be understood. As the elements of a uni-

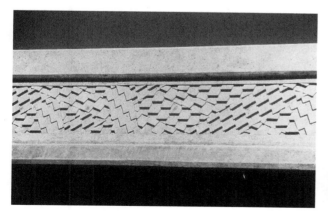

FIGURE 5-47 *This pattern creates random patterns of shadows that change at every sun angle.*

FIGURE 5-48 *A vertical pattern of closely spaced ribs hides drainage stains and construction irregularities.*

form pattern become smaller, they are not seen as individual elements but as components of a texture. Textures will be seen primarily as color modifiers.

Texture does become more important at street and pedestrian speeds. People have the time to see the subtler features—the mortar lines in brick surfaces, for example—and even reach out and touch surfaces that are inviting. Even at this level, however, the color of the surface and its relation to surrounding colors must be understood independent of the texture.

Concrete, whether used as a girder or wall, can be patterned in many ways, limited only by the imagination of the designer and the size of the budget. A pattern can be as simple as a pattern of incisions based around standard form panel dimensions and concrete lifts. Possibilities beyond that include raised or recessed panels, ribs, and indentations (Figure 5-47 and 5-48).

Traditional materials such as brick and stone have their own characteristic patterns and textures. They also have associations that we carry over from their use in architecture. A particular type and color of brick, for example, may remind us of the buildings of a certain town. To the extent that these associations are widely shared in a community, the use of that material may carry a message to the larger population that associates the bridge with a certain historical period or social values associated with that period. For example, using concrete patterned to look like the stone of the state capitol building can give a nearby bridge a sense of dignity because of its association with the capitol. (See Figure 3-13, page 77.)

FIGURE 5-49 *Pima Freeway, Phoenix, Arizona. A figurative pattern on an abutment can work well if it is contained within the main lines of the structure.*

Concrete Patterns and Textures

Concrete offers many possibilities for the imposition of various types of pattern through form liners, custom formwork, and other devices. To a certain extent, the designer has no choice but to create some level of pattern in large concrete surfaces. The form joints and construction joints of the concrete will create a pattern, whether the designer recognizes it or not. The result is likely to be much improved if the designer at least controls these patterns through design, rather than leave them to the vagaries of the construction process. The temptation to add variety needs to be disciplined by the design intention so that patterns do not disrupt the forms of the structural members.

Subordinate Patterns on Bridges to the Design Intention and the Forms of the Structural Features

It is usually best to avoid patterns on girders. These elements make their impression by their overall shape and the appearance of strength they convey. Patterns weaken that effect. The only concern on girders should be controlling the lines created by stiffeners, construction, and form joints so that they are consistent with the overall shape of the girder and occur at controlled intervals. Figure 5-36 (page 162) is an example of a poorly chosen pattern for a girder.

Similarly, patterns on piers can interfere with the appearance of strength if they appear to interfere with the logical flow of forces. Again, the first goal should be to organize the construction and form joints so that they are consistent with the overall shape of the pier.

Abutment walls offer more possibilities for pattern and ornamentation. For most highway bridges, focusing on the wing walls is the more effective use of resources (Figure 5-49), as the abutment face is not easily visible to drivers passing under the bridge. However, if there is significant pedestrian traffic on the undercrossing roadway, walkers will appreciate attention paid to the abutment face. Any pattern or ornamentation used should respect the overall shape of the abutment elements. Wing walls are usually triangular, and a pattern or ornamental element needs to fit this space. The pattern also needs to be consistent with railing designs and other elements of the structure.

Any Feature Placed on the Abutment Should Help Unify the Structure, Not Make the Abutment Appear as a Separate Element

An inappropriate feature on the abutment can distract attention from the structure as a whole. When developing abutment surface treatments:

- Lay out the elevation of an entire abutment and any adjoining retaining wall at one time.

- Recognize and organize expansion, contraction, and construction joints.

- Ensure that all lines bear some obvious relation to the main lines of the structure.

Organize Retaining Wall Surfaces

To accomplish this, follow these guidelines:

- Organize the expansion, contraction, and construction joints. Drain holes at the base of the wall should be installed in a consistent and logical relationship to the surface pattern and, if possible, be hidden in the wall pattern.

- Pattern lines should parallel the grade of the lower roadway, or be horizontal or vertical. Lines that are close to but not identical to the roadway geometry and other features will be particularly noticeable and jarring. Continuous horizontal lines should be either level or follow the major lines of the roadway. Horizontal lines must be carefully controlled, as any irregularities will be immediately obvious. Textures created by repeated linear elements (such as ribs) are more successful when they are primarily made up of vertical elements, because construction control is not so critical.

- Control frequency of repeating patterns. Patterns that repeat abruptly and continuously at intervals of about one-quarter second to one second at the prevailing speed (20 to 75 feet at 50 mph) will be annoying and quickly become monotonous (the "telephone pole" effect). Patterns at shorter intervals will become a textural blur. Longer patterns (200 feet-plus) will be seen as adding variety, unless they are repeated too often without change.

- Create patterns consisting of elements that are large enough to be seen at highway speeds. The minimum dimension is approximately 4 inches. Elongate pattern designs horizontally to compensate for perspective foreshortening.

FIGURE 5-50 *Squaw Valley Parkway, Phoenix, Arizona. The pattern of this retaining wall plus noise wall has repetitive elements so that it can be coordinated with form panels and enough detail to be interesting, even to motorists stalled in a traffic jam; but its overall emphasis is a horizontality that reinforces the lines of the highway and hides the joint between the retaining wall and noise wall.*

FIGURE 5-51 *Striking patterns can be made simply by alternating areas of smooth concrete with area of texture created by form liners. On this wall on I-71 in Columbus, Ohio, the textured area is created by evenly spaced small vertical ribs.*

One point is clear: studying the pattern on elevation drawings is misleading if the wall will be seen primarily from a moving car. The wall will actually be seen as if it were on a vertical scroll unwinding in the peripheral vision. The impression created will bear little relation to what can be gleaned from the elevation drawing.

When Working with Walls Made Up of Repetitive Precast Facing Units, Develop a Consistent Strategy for Dealing with the Wall Pattern

Walls made up of repetitive precast units have specific visual characteristics. The units themselves will create a unique pattern that is large enough to be read in the highway environment. The patterns often end in jagged hexagonal or cruciform edges. This pattern can be made more noticeable by creating protrusions, recesses, or exaggerated edges in the precast units. Or, it can be somewhat concealed by overlaying regular patterns created by form liners.

FIGURE 5-52 *Sometimes the best pattern is the continuation of natural vegetation down to roadway level.*

- Develop a pattern that works for all suppliers. Bridge-building agencies sometimes require alternative bidding by manufacturers that supply different unit shapes. A design predicated on a hexagonal unit may be a different thing entirely if applied to a cruciform unit. If alternate bidders must be accommodated, the challenge will be to develop a design that applies equally well to all potential bidders. Alternatively, if a specific appearance is particularly important, bids should be restricted to manufacturers offering compatible units.

- Develop and apply a consistent method of completing the pattern at its ends. In the "Retaining Walls and Abutment Wing Walls" section (page 161), a poured-in-place concrete cap is suggested to provide a smooth ending at the top. A similar strategy should be followed at the ends unless the pattern can be made to turn a corner. The joint with a cast-in-place wall should be a simple vertical element that clearly ends the pattern of the precast elements and allows another pattern to begin. A joint that is a reciprocal shape to the precast elements usually looks like a poor attempt at imitation.

- Where precast unit walls adjoin poured walls, take care that the pattern and color of the poured wall is compatible with those of the precast wall. This is not to say that the cast-in-place wall should be scored to imitate joints of the precast facing units. The result will not be convincing. Better to contrast a larger pattern, taking care that the colors are compatible.

- For transitions between precast retaining wall sections and a cast-in-place retaining wall or abutment, use a simple cast-in-place vertical feature at a logical location (perhaps at the expansion joint). Surface treatment of abutments should be consistent with parapet treatments and pier design. For most structures, the only abutment surfaces clearly visible will be the side walls. Face walls are too foreshortened and usually too shadowed to be worth special treatment. Exceptions are unusually wide structures, skewed structures, and bridges crossing city streets or pedestrian areas.

FIGURE 5-53.1 *Randomly spaced vertical ribs break up the pattern of the MSE wall panels and carry over to the cast-in-place abutment to create continuity.*

FIGURE 5-53.2 *The pattern on this long wall has enough random elements in it to disguise its repetition.*

Select Form Liners That Support the Design Intention
and Are Consistent with Nearby Features

Rubber and plastic form liners are an economical way to achieve interest and texture. What clues does the immediate area offer? Brick buildings? Fieldstone walls? Woods? What will coordinate with other nearby highway structures? Surfaces can be chosen that blend in or contrast, whichever fits the overall concept (Figures 5-53.1 and 5-53.2).

• In the highway environment, use form liners that are abstract or random patterns with elements large enough to be recognizable. Long lengths of retaining wall or noise wall next to a roadway are seen with so much perspective foreshortening that the repetitive pattern of the liner becomes obvious. The result is a large-scale version of the boredom created in a large room with walls covered by wallpaper with a small repeating pattern. Random and irregular patterns hide the repetition better. They also are more consistent with the random textures in nature (tree bark, rock outcrops), which are frequently the background. Vertical grooves at random intervals create a simple pattern that works well. Horizontal patterns should be used with care because they are visually very prominent, and any flaw in their alignment will be glaringly obvious. Similarly, diagonal and curved patterns should be used with care, as they will appear to swoop and dive as the viewer passes them at highway speeds, with a potentially disconcerting effect.

• Be aware that form liners that seek to imitate brick or stone are unconvincing unless special care is taken. The effect can work only if the effort is made to color each of the individual "stones" or "bricks" individually and differentially, and to hide the joints between panels. To be convincing, the liner also has to provide enough physical relief to match the depth of the mortar joints and stone texture found in real masonry. The best result is gained when the form liner itself is a mold taken from an actual masonry wall.

• For most highway bridges, ensure that form liners or surface treatments that attempt to imitate random wood boarding have elements large enough to be seen at highway speed. Smaller elements may be applicable in areas where pedestrians circulate. The wall shown in Figure 5-23 (page 155) is a good use of a large-scale, random form liner of vertical elements. Surfaces in pedestrian environments offer a wider set of possibilities because pedestrians are close enough, and moving slowly enough, to appreciate them. Decisions about them should be made based on the prior development of the deign intention for the structure. What is the predominant use of the area? Waiting for a bus? Walking to school? Sitting in the sun? How can the highway contribute to that use? Or protect itself? Is the highway intended to be friendly or unfriendly? For example, a surface can be created by fracturing protruding vertical concrete ribs with a hammer, which produces myriad sharp edges. Such a surface would be a bad choice adjacent to a playground, but might be a good choice to discourage loitering in another area.

Surface treatments that break down the concrete surface (e.g., bush hammering, acid wash) are not generally successful in highway environments. They can not be "read" at highway speeds, and the surface becomes more porous and more susceptible to dirt and deterioration.

Facing Materials

Use Traditional Masonry Materials in Ways Consistent with Their Structural Capabilities and Historic Use

Masonry materials find their most logical application for facing of abutments and retaining walls, circumstances in which they have been used structurally for centuries (Figure 5-54; Figure 3-15.1, page 78). They are not convincing as part of a span unless they are formed into a structural arch, consistent with their historic use. If they are used as part of a parapet, they need to be detailed in such a manner that the support provided by the girder below is clearly demonstrated. Otherwise, the viewer will have the uneasy feeling that they are unsupported, and may fall, as in Figure 3-15.2 (page 78).

Of the various facing materials, random fieldstone with large units (greater than 18 inches) appears to work best, as its size and texture make it visible in the highway environment. Precast concrete panels can work well, also. The major concerns are surface color and texture and panel joints. Brick, with its mortar pattern, is on the small side for highway uses, but can still add interest. It is more appropriate to a pedestrian environment.

FIGURE 5-54 *The Starruca Viaduct, in northeastern Pennsylvania, is an excellent example of large stone masonry from the nineteenth century. Its form also perfectly illustrates the structurally honest use of stone in a bridge.*

Before Making a Final Decision on Color/Texture of Masonry or Precast Concrete Facing Materials, Construct Sizable Sample Panels On-Site to View under Various Light Conditions

Color selection is even more critical with masonry materials, as the possibility of a future change is remote. Brick must be handled especially carefully, as the range of color choices is high, and the possibilities for an inappropriate choice multiply. With brick and stone, mortar color must also be considered.

Consider Metal Panel Systems to Provide a Ceiling for Pedestrian Areas below Bridges

Many structural systems, particularly steel and precast concrete I-girders, do not have an attractive underside. If pedestrians will be using the space beneath, metal panel systems can provide a clean attractive surface that reflects light into the underbridge area. Durable, economical systems can be made from aluminum, stainless-steel screening, and fiber-reinforced plastic.

Ornamentation

Historical structures, especially those located in a city center or other important spot, were often elaborated with decorative railings and detail. The beginning and end of a bridge were often recognized with pylons or statuary. These features added interest and visually recognized the symbolic role of the structure and its position within the larger urban context. Citizens often feel that unadorned modern structures insufficiently acknowledge these needs.

Because aesthetic quality in bridges results from the perfection of engineered form, the first priority of the engineer should be to shape the structure so that the structural elements themselves produce the intended aesthetic impact. Well-designed structural details can articulate and emphasize the structural shape. The fittings of cable-supported bridges, for example, can make a big impact on the quality of the aesthetic impression (see Figure 6-83, page 276). Chapter Four discussed arranging stiffeners on steel girders to create a natural emphasis on the bearing point in a way that might be called ornamental, and still serve the structural function. Indeed, many of the historical systems of architectural ornamentation had their beginnings in the elaboration of structural elements. However, most of these systems evolved in masonry buildings, and had their beginnings in the structural strengths and weaknesses of stone and brick. Imitating these forms in modern materials makes no sense at all, and will always appear out of place. In contrast, the art deco ornamentation of the Golden Gate Bridge (see Figure 1-2, page 2) depends on geometric shapes easily formed in steel and concrete, which are consistent with, and stay within, the overall shapes of the structural members.

If not designed carefully, ornamentation and nonstructural surface materials can disguise, detract from, or destroy the structural form. This was the general practice in late-nineteenth-century architecture, and was the reason that the first modern architects looked to nineteenth-century engineering as an example of functional clarity and pure form. Unfortunately, the earlier practice has revived for bridges in historic areas. In their efforts to re-create the visual interest of a beloved historical bridge that must be replaced, or just to improve the appearance of an otherwise ordinary structure, designers are adding elements derived from historic architectural styles to bridges. The result has been bridges in which the lines of the structure are blocked or hidden with ill-proportioned, garish, and/or tacky details (Figure 5-55).

Modern values in any case focus on functionality, and elements are expected to

FIGURE 5-55 *The traditional pilasters and railing added late in the design of this otherwise ordinary precast girder bridge have no relationship to the underlying structure; they also block its lines and make it appear top-heavy.*

FIGURE 5-56 *The railing on the 90th Street Bridge in Redmond, Washington, is consistent with the contemporary structural system; it reinforces the lines of the structure and adds design interest for pedestrians.*

have some use beyond creating a visual impact. Thus, many successful contemporary ornamental systems elaborate the design of functional features, such as railings and lighting support, in ways that reinforce the lines of the structural members (Figure 5-56).

Ornamentation can add additional levels of interest and richness, but it is best when restricted to those locations with a commensurate level of importance and exposure, and when it reflects the forms and abilities of modern structural materials. The critical criterion is to make sure that the ornamentation emphasizes, rather than camouflages, the structural form.

Details

Bracing, scuppers, drain pipes, conduits, and utilities are often more visible than most engineers realize. These functional necessities are not usually thought of when the design intention for the bridge is initially conceived. However, they creep onto the structure in the final stages of design and sometimes overwhelm its best features. These details will be particularly visible wherever pedestrians are nearby. The higher the structure, the more visible the details. It can have a much greater impact on the final appearance of the structure than can be predicted from elevation or section drawings. The result can undermine the best efforts of the designer on other elements of the structure. As one bridge neighbor was heard to say, "From most angles the bridge looks fine, but from below it looks like the underside of an old car." The goal should be to reduce the amount of such features, simplify those that remain, and design them so that they generally follow the main lines of the structure.

Design Bracing to Be Seen

Simplification is the key. Patterns with few members arranged in simple and consistent configurations should be used. (See 5-57 and also Figure 6-26.2, page

FIGURE 5-57 *Simplified bracing patterns will improve the appearance of the underside.*

217). Complex patterns distract the eye and interfere with any positive effect created by the main structural members.

- Use patterns with all diagonal or all right-angled members.
- If diagonal bracing is used, keep all members at the same angle relative to the superstructure.
- Use a few large bracing members rather than many small members.

Where Possible, Eliminate Drainage Inlets on Bridges

The best answer is to just not have drainage inlets on the bridge. This can often be achieved on shorter bridges by adjustments to the vertical geometry to eliminate sumps on the bridge, changes to the shoulder cross slope to allow it to carry water further, and other relatively minor changes. Placing inlets on the roadway at the ends of the bridge is one obvious strategy.

If Scuppers Are Absolutely Necessary, Avoid Piping Systems

Recent water quality restrictions have reduced the applicability of this approach, but it is still the best idea. Drainage pipes, in addition to their visual drawbacks, are notorious maintenance problems.

- Look first for a location where water can be permitted to outlet just below the bottom of the girder. For bridges over bodies of water, this can be almost anywhere in the bridge. For bridges over land, potential locations must fit into drainage patterns of the area below. A splash block below the outlet will usually be required.
- Design scuppers so the outlet falls on the inside of the fascia girder and still accommodates the desired overhang. With scuppers that outlet through the parapet, integrate the outlet design with the design of the parapet. This will place restrictions on scupper locations because, as visible elements, their location must be coordinated with the overall concept of the bridge.

If Drain Pipes Must Be Installed, Hide Them in Spaces between Girders and within Piers

Pipe systems embedded within piers should be avoided in cold climates because of freezing problems. An alternative is to place the drain in a groove in the pier where it seems to be part of the pier but is still accessible for maintenance.

If Drainage Systems Must Be Exposed, Design Them to Be Seen

It does no good to pretend the drainage system will not be seen or to leave its design to the contractor's pipe fitter. Better to consciously design the system to be a complement to the bridge itself (Figure 5-58).

FIGURE 5-58 *Piping at this structure for Washington D.C.'s Metro is designed to be seen, with parallel large radius curves and few fittings.*

- Keep pipe systems as simple as possible, using large-radius curves and as few fittings as possible. For example, it is better to

FIGURE 5-59 *An example of how signing, lighting, and landscaping can be successfully integrated into a bridge. This is the same bridge as in Figure 2-42 but shown before the planting matured.*

lay a pipe in one simple run down the face of a column rather than install five elbows to get it around to the rear.

- Wherever drainpipes are exposed, run them in lines that are either parallel or at right angles to the main lines of the structure.

- Paint drain pipes the same color as the structural element against which they are mounted.

Conduits and utilities impose similar problems as drainage systems, and should be approached using the same basic ideas. These ideas will not eliminate visual problems with drainage, utilities, and other piping, but they will at least help make these elements a more positive part of the structure.

FIGURE 5-60 *A sign on a simple independent sign structure is almost always preferable to mounting a sign on a bridge, even if the sign occurs just a short distance ahead of the bridge. This sign would look better if it were a simple rectangle rather than an irregular six-sided figure. Compare with the sign in Figure 5-61. The money saved by reducing the size of the upper panel is negligible.*

FIGURE 5-61 *Curvilinear pipe sign structures reduce the visual presence of this utilitarian and desirable background structure to a minimum. Coordinating the sizes of adjoining signs and simplifying their shapes also improves the structure's appearance.*

FIGURE 5-62.1 AND 5-62.2

Signs on bridges should be simplified in shape and related in size to the bridge.

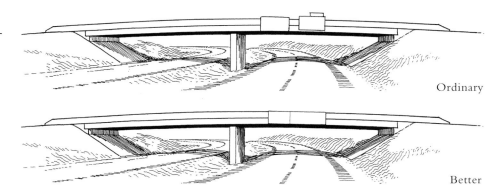

Ordinary

Better

SIGNING, LIGHTING, AND LANDSCAPING

While these elements are not, strictly speaking, part of the bridge itself, they can have major impact on the structure's appearance. The goal is to include them within the design intention and integrate them into the design of the bridge (Figure 5-59).

Signing

There are two types of signs on structures. The first and most common is where the bridge itself is used as a support for a sign serving the underpassing roadway. The second is when a sign structure is erected on a bridge to serve the bridge's own roadway, which is often necessary on long viaducts and ramps. In both types the sign usually blocks and/or complicates the lines of the bridge itself. The result is rarely attractive. Thus, the most desirable option is to keep signing off of structures. It does not do any good to create an attractive bridge design and then saddle it with an ugly sign or sign bridge. The first goal should be to seek an alternative location for signs away from a structure. This will inevitably mean more specialized structures for the signs themselves. A simple, attractive sign structure even 300 feet *before* a bridge is preferable to a sign *on* the bridge itself (Figures 5-60 and 5-61). However, there are situations in which the highway layout is so constrained that the only reasonable location for a sign is closer than 300 feet. In those cases, the driver's view through the sign bridge should be checked. If the sign bridge effectively blocks the view of the roadway bridge, the sign might as well be on the roadway bridge.

When a Bridge-Mounted Sign Is Required for the Underpassing Roadway, Fit Sign(s) into the Overall Design

Align the top of the sign with the top of the parapet, railing, or pedestrian fence, and align the bottom of the sign with the bottom of the superstructure (Figure 5-62.2). This may require a nonstandard arrangement of the message. Where more than one sign is on a bridge, all of the vertical dimensions of the signs should be the same.

Keep Sign Bridges Simple

As discussed in Chapter One, one of the secrets of success in highway and bridge aesthetics is to keep everything as simple as possible. Sign bridges anywhere, on structure or off, look best when they are simple. These structures should be simple and

FIGURE 5-63 *Sign structures on bridges should be coordinated with the structure; for example, by being placed on a pier.*

cleanly designed, relying on a few large members rather than truss work (Figure 5-61, page 187). It is hard to get simpler than a bridge made of a single curved tube. Even the simplest trussed sign bridge counts as three elements: the two posts and the truss that doesn't even include all of the individual members of the truss. Structures using two tubes with Virendiehl bracing can be designed to carry signs too large for a single tube.

> ### Design Connections for Sign Structures on Bridges as a Logical Extension of the Structural Members of the Bridge

When a sign structure must be mounted on a bridge structure, the connection should be made so that the sign structure looks like it belongs there, not like a slapped-on afterthought (Figure 5-63).

Lighting

There are two areas of concern here. The first, and more common, is the mounting of light poles and fixtures on a bridge to light the roadway and/or the area blow. The second is lighting the bridge itself.

Usually the pattern of roadway lighting depends on a regular, even spacing of poles. The pattern frequently begins some distance from the bridge, and the poles end up at odd locations on the bridge, placed wherever the pattern dictated, sometimes with unfortunate consequences for the appearance of the bridge. Roadway lighting fixtures are not that precise. Small adjustments in spacing can be made with-

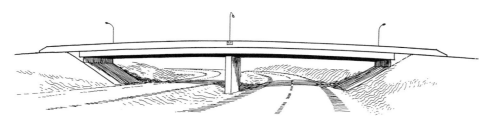

FIGURE 5-64 *Lighting should be organized around structural features.*

FIGURE 5-65 *The support area for the light mounted on the bridge is continuous with the parapet; however, placing it over the pier would have been an improvement.*

out appreciably affecting the uniformity of light levels. Such adjustments should be made to bring the poles into a better visual relationship to the bridge.

- For short bridges, space poles on adjoining roadway lighting to avoid the need for a pole on the bridge.

- If lighting poles on the bridge cannot be avoided, place roadway lighting poles in some relationship to the structural features.

Lighting poles should be positioned so as not to distract from the overall design. One way to do that is to have a consistent relationship with the pier spacing, by, for example, placing each pole at a pier (Figure 5-64).

Mount Lighting Poles on a Widened Area in the Parapet

The support element design needs to be as continuous as possible with the parapet itself and consistent with any railing posts, grooves, or recesses and construction joints of the parapet. The design goal should be to maintain the horizontal line of the parapet with as little interruption as possible (Figure 5-65).

For Lights under Structures, Maintain Some Relationship to the Structural Features, to Emphasize the Structural Form

For example, placing lights between every third girder of a girder bridge would not only light the roadway below, but light every third space between the girders, creating a rhythmic interval of light across the "ceiling" of the space underneath the bridge. Light fixtures mounted under a bridge to light the underpassing roadway offer a special opportunity to make the bridge a nighttime feature. The area beneath the bridge usually becomes the brightest part of the nighttime visual field, creating a "lighted portal" effect.

Consider Lighting the Bridge Itself

There are specific instances where, because of the size of the bridge itself, its location in the community or the environment, or its symbolic importance, the bridge itself deserves lighting. There are three basic approaches in this situation:

FIGURE 5-66 *Broadway Bridge, Miami, Florida. Floodlights reinforce the shape of the structural elements (see color insert).*

- To simply floodlight the bridge. This will not replicate daytime appearance, because the shadow areas and the color effects will be unavoidably different. However, it does come closest to giving a complete picture of the bridge.

- To outline significant features of the bridge in individual lights. The lights produce a row of lighted dots that trace the silhouette of the structure. Variations include having colored lights or having the individual lights go on and off in some pattern.

- To light only significant portions of the bridge. This works well when it creates a pattern that is essentially an abstraction of the basic structural form. An example is lighting the cables of a cable-stayed bridge. Lighting the supports and underside of the bridge also works well (Figure 5-66). It turns the bridge into a series of lighted portals and illuminates the feature the bridge is crossing, while minimizing spillover light.

Development of an aesthetic lighting concept should begin by addressing the night-time appearance of the bridge as part of the original design intention/vision to ensure that the lighting concept is consistent with the wishes of the community and the structural concept of the bridge. Successful aesthetic bridge lighting requires a great deal of specialized expertise and experience. Interested designers should consult with lighting design consultants to produce the actual design.

Landscaping

Landscaping is defined here to include planted areas and hardscape: stone, brick, or concrete paving, often colored and/or patterned, used primarily for erosion control or pedestrian circulation.

Landscaping should enhance an already attractive structure. It should not be relied upon to cover up an embarrassment or hide some unfortunate detail. Conversely, it should not be allowed to grow up to hide some important feature that is crucial to the visual form of the bridge. Landscaping can be a more economical and

effective way to add richness and interest to a design than special surface finishes or materials. For example, a large, plain concrete abutment can be effectively enhanced by well-chosen landscaping (Figure 5-67).

The application of landscape concerns, such as environmental suitability, topography, and existing vegetation, should be part of the site analysis and development of the design intention/vision for the bridge as a whole. The result should be, in addition to a design intention for the bridge, a consistent design intention for adjacent landscaping. Landscaping concepts developed for a larger project should recognize the design intention for the structures on that route.

Landscape planting must be consistent with the environment. Plants that are native to an area, or at least grow well in the area without irrigation or other special treatment, work best and will seem most consistent with their surroundings. Selecting plants for these purposes requires detailed specialized knowledge and should be done in collaboration with a local landscape architect. The landscape architect should be a member of the team from the beginning of the concept development for any project in which landscaping is expected to be a significant component.

Use Planting to Enrich the Appearance of Abutments

Planting can blur the division between environment and bridge, tie the bridge into its surroundings, and add visual interest to what can be massive, bland elements.

In Rural Areas, Use Landscaping to Emphasize the Continuity of Space through the Bridge

If the design intention of the bridge is to ensure the maximum openness of the space under the bridge, as well as an unobstructed flow of space through the bridge,

FIGURE 5-68 *Continuity of landscape patterns between foreground and background results in a pleasing uninterrupted vista.*

FIGURE 5-69 *Extensions of the abutment wing walls as terraces integrate a bridge into an overall interchange design.*

landscape planting can help achieve this effect. A recognizable pattern of planting can be started before the bridge, continued right through it (with a minimal interruption directly underneath), and for some distance beyond (Figure 5-68).

Use Landscaping to Mark a Milestone

The infield areas of interchanges and the outer areas of rights-of-way can be used for significant landscaping which, together with the bridge(s), mark the location as a milestone (Figure 5-69).

Use Landscaping to Restore the Space below a Structure

Areas below viaducts can become wastelands of blowing wastepaper and abandoned automobiles. Landscaping restores the attractiveness of such areas and signals that someone cares about the space, thus reducing its vulnerability to vandalism.

FIGURE 5-70 *The ground plane under a viaduct should be restored to an attractive condition through the use of attractive hardscape.*

FIGURE 5-71 *Slope paving carried to an extreme: These armored slopes seem part of the structure, adding two large visual masses that weigh down an otherwise graceful bridge.*

Plantings can be used in those areas that receive sufficient light and moisture, and hardscape used elsewhere (Figure 5-70).

In More Urban Areas, Use Landscaping to Relieve and Contrast with the Hard-Edged Elements in the Visual Field

The buildup of many different structural features in urban areas—bridges, retaining walls, and noise walls—often becomes overwhelming. Landscaping can give back some green to neighbors and road users. As Figure 2-55 (page 57) shows, small auxiliary walls can be placed in front of major walls or abutments to create level

FIGURE 5-72 *Riprap allows plants to blur the edge of the slope paving and blend into the landscape. Riprap does a better job of slope protection as well, replacing undermined concrete slabs in this example.*

planting areas. In locations where a bridge adjoins a community or group of buildings, landscaping can be an indispensable element in mediating the differences in material and scale. Here, the goal is probably best served by intensifying planting patterns and species already existing in the community, in order to emphasize the continuity of the community environment.

Pick Landscaping Colors and Shapes in Relation to Those of the Bridge

Landscaping creates its own shapes and colors. These must be related to the shapes and colors of the bridge. Even though highway landscaping generally works best and is most easily maintained when the materials and patterns replicate the existing natural vegetation in the immediate area, there will always be locations where contrasting plantings are desired in order to accentuate or emphasize a particular bridge location or structural form. As long as the choice is consistent with the design intention of the structure itself and the design theme of the highway, this can be legitimate in the same way that it is legitimate to paint a bridge a color that contrasts with its environment. Needless to say, the resources and commitment to maintain such plantings must not be in doubt.

Use "Landscape" Material for Slope Protection

Slope protection is primarily a landscaping job, the function of which is erosion control. However, if concrete is used, the slope protection appears to be a structural material that is part of the bridge. The concrete slopes thus make the bridge appear larger and heavier (Figure 5-71). Slope protection should be performed by a landscaping material, such as riprap stone, placed so it looks like part of the landscape. Riprap has the advantage that plants will tend to colonize it and blur the edges with vegetation (Figure 5-72).

Use only enough slope protection to cover the area beneath the bridge where plants cannot grow. Drainage channels from the bridge should be accommodated with separate riprap channels or piping. Dark-colored stone is better for riprap, since it will attract less attention. If concrete must be used, try to achieve a random pattern in the surface. Patterns pressed into newly placed concrete can give a good appearance at minimum cost and break up the flat surface of conventional concrete for paving. Colors should be different from and darker than the structure colors.

---

1   Holtzschue, Linda, 2002. *Understanding Color,* New York: John Wiley & Sons, Inc.

2   Lambert, Patricia, 1991. *Controlling Color,* New York: Design Press/McGraw-Hill, Inc.

3   The standard assortment of colored pencils and markers found in an office or technical supply store won't do, either. They are too restricted as to shade and hue, and don't correctly represent the colors found in nature. Shop at an art supply store. Look for Berol Prismacolor pencils and markers and AD art markers or their equivalents.

4   Zuk, William, 1974. "Public Response to Bridge Colors," *Transportation Research Record*. Washington, DC: Transportation Research Board, vol. 507.

5   Porter, Tom, and Mikellides, Bryon, 1976. *Color for Architecture.* New York: Van Nostrand Reinhold.

6   Pantone Color Matching Systems, Pantone, Inc., 55 Knickerbocker Road, Moonachie, NJ.

chapter **six**

PUTTING IT ALL TOGETHER: TYPICAL BRIDGES

"The greatest glory in the art of building is to have a good sense of what is appropriate. For to build is a matter of necessity; to build conveniently is the product of both necessity and utility; but to build something praised by the munificent, yet not rejected by the frugal, is the province of an artist of experience, wisdom, and thoughtful deliberation."

—Leon Battista Alberti, *On the Art of Building*, 1486

Chapter Three discussed the process of bridge design, beginning with the development of a design intention/vision for each bridge. Chapters Four and Five developed a design language for the various bridge elements in the order of their importance as Determinants of Appearance. The components of the language are a series of guidelines for the design of the individual elements that will help to produce high-quality appearance in bridges. This chapter shows how to put all of these ideas together, using typical bridge sites as examples.

Developing a design for an actual bridge site will follow a similar pattern to the typical sites discussed in this chapter, but an actual site will have specific features that may lead to a different result than the examples given here. The designer must be aware of these differences and follow where they lead. Mere imitation of the examples given here will not lead automatically to success, and may do just the opposite. After all, it is that exceptional match of need and solution that provides most satisfaction in life, in bridge design as in many other areas of endeavor.

The discussion of each bridge type begins with the development of a design intention/vision, as described in Chapter Three, and is followed by an illustration of how to apply the Determinants of Appearance to that bridge type.

HIGHWAY-OVER-HIGHWAY BRIDGES

This section focuses on the typical highway-over-highway bridge of one to four spans, where both ends of the structure can be seen at one time. Longer structures are classified as ramps and viaducts and are discussed in the "Interchange Ramps and Viaducts" section, later in this chapter.

Developing a Design Intention/Vision

Physical Requirements

The site requirements are largely determined by the geometry of both roadways and horizontal and vertical clearance zones of the undercrossing roadways. Occasionally overlooked is the desirability of providing sight distance through the structure to objects that may lie beyond, such as a ramp nose or traffic signal. Utilities associated with either the undercrossing or overcrossing roadway are sometimes a consideration. Geotechnical considerations affecting potential foundations must also be understood. The typical requirements will result in structures of one to four spans with spans in the range of 50 to 200 feet.

Visual Environment

Bridges associated with freeways and major highways will usually be most affected by the environment created by the highway itself: its slopes, landscaping, retaining walls, and noise walls. In rural and suburban areas, this environment is typically wide open with extensive horizontal vistas. The predominant influence will be the landscaping/cover of the ground surface and any retaining or noise wall in the scene. The structure will most often be seen against a background of sky.

In very mountainous areas and urban areas, the mountainsides, walls, and any visible buildings will limit the vistas. Then the background will be heavily influenced by the materials of the mountainside and any vertical surfaces.

The most important viewpoints for highway bridges are usually points along the undercrossing roadway. Because of the way people see things when moving at highway speeds, bridge crossings over freeways and expressways are seen best when they are still distant. At highway speeds, the closest point at which such a bridge will "register" to a motorist is 300 to 500 feet away, depending on the actual speed. At any point closer, the motorist is focusing beyond the bridge, and the bridge itself is a blur in his or her peripheral vision. Because bridges over such highways are seen by people who are moving along predictable paths, at predictable speeds, it is easier to predict what people will see at each point and to control what their perception will be.

Bridges over arterial streets are viewed from more varied locations. A survey of the surrounding area will indicate the important viewpoints. The designer should particularly look for places where people congregate, such as the entrance of a nearby shopping center, or the stands of an adjacent baseball diamond.

Nearby and Associated Uses

Bridges over arterial streets are often closer to existing or planned land uses, such as retail or residential areas, with associated buildings. The buildings become part of the backdrop for the structure. Their visual effect on the structure, and its effect on them, must be considered. Sometimes the structure will be considered an important feature in an urban district, and nearby building owners will want attention paid to its compatibility with existing buildings.

Often structures in this situation must provide for pedestrian use as well. The nature of this use, children going to and from school, for example, should be considered. If pedestrians will be present, clear visibility to all parts of the pedestrian walkway will be important.

Symbolic Functions

Most highway overcrossings are subordinate parts of larger transportation facilities. Their symbolic role, to the extent that they have one, is to be seen as enabling a clear and safe passage. The message should be one of free and continuous movement. Occasionally the position of a structure within a town or neighborhood may lead to a requirement that the structure be seen as a gateway or that it acknowledge nearby architectural styles.

Boundaries

Typically, the whole structure can be seen at a glance, so it is important to set boundaries that are logical and obvious. The structure and its abutments, including any attached retaining walls, should be seen as one unit out to the end of the parapets. Guardrails are seen more as site features than structural features, but their connection to the parapet will be apparent and should be done as a clear and simple transition.

Guidelines for the Design Intention/Vision

At distances of 300 to 500 feet or more, the only parts of the bridge clearly visible are the features of the front elevation. This means that the features of the elevation of the bridge will determine the visual impression. The most important of these features is the face of the superstructure, followed by the face of the parapet and any railings or pedestrian fences. Next in importance are the end elevation of the pier, the wing walls of the abutments, and the shapes of the voids between these elements. Finally, the method of ending the parapet features will also have an influence.

Exceptions exist for bridges on a severe skew, very wide bridges, or bridges crossing a sharply curved or steeply sloped roadway. In these cases, portions of the abutment face or the side of the pier may become important components of the visual field, unless they are in shadow. Sharply curved, highly superelevated bridges may expose their undersides to view. These situations need to be analyzed on a case-by-case basis, using perspective views taken from viewer locations on the undercrossing roadway. If pedestrians use the undercrossing street, the appearance of the underside will be important.

Most highway bridges are short. Usually the driver experiences them as relatively minor elements along a much longer path. The view through them to the path beyond is almost always critical. It is important that they be kept simple, with all parts in clear relationship to one another and an overall appearance of unity.

When designing a highway-over-highway bridge:

- Keep things simple, with the fewest number of different elements and shapes.

- Provide generous vistas through the structure.

- Develop apparent thinness in structural members.

- Choose shapes from the same family. Faceted piers should be used with faceted parapet design; rounded pier designs with rounded parapets.

- Use a minimum number of different materials, different colors, and different textures.

- Always use a given material, color, or texture the same way.

Highway overcrossings often come in groups; for example, all of the bridges on a section of freeway, or all of the bridges in an interchange. Because multiple bridges will be seen in quick succession, it is necessary to consider their relation to each other, so that the result is not a visual hodgepodge. (See Figure 6-16, page 207, and the discussion in the "Design Themes" section in Chapter Three.)

Determinants of Appearance

Geometry

The geometry of a highway overcrossing is largely determined by the needs of the intersecting roadways. That said, there may be room for adjustments, and this should be investigated.

The structure will generally look better if it is on a vertical crest curve (Figure 6-1). This can often be accomplished without lengthening the structure, though it may require slightly higher piers. Sag vertical curves and horizontal curves may be acceptable if they are long enough. Make sure any alignment curves are at least three-quarters the length of the structure; otherwise, the structure will appear to contain a kink, creating a visual problem that cannot be resolved by later decisions.

Superstructure Type

Because the elevation view is very important in the appearance of this type of structure, the proportion and relative slenderness of the girder will have a major effect in forming people's impression of the structure. Relative slenderness should be sought when selecting superstructure type. Haunched girders can create an impression of slenderness, as well as tell a story about how the structure works, thereby adding an element of interest. The haunched girder with minimal abutments in Figure 6-2 completely opens up the view through the bridge. The open railing and the narrowing of the pier at its base make the structure seem light. The haunch shows how the bridge works, and the stiffeners draw attention to the point-of-force transfer. There are two unfortunate elements in the picture, however: The concrete slope pavement creates a visual mass that weighs the bridge down at its ends, and there should have been another location for the sign that notifies travelers of an interchange one mile ahead.

FIGURE 6-1 *The crest curve creates a slight arch that most people find attractive; it also appears to open up the view through the bridge.*

FIGURE 6-2 *This haunched girder with minimal abutments completely opens up the view through the bridge. The unnecessarily extensive slope pairing makes the ends of the bridge appear unfortunately massive.*

Steel and concrete rigid frames were popular for highway overcrossings in the early days of freeway building, when span lengths were generally shorter. They looked good because the midspan depth could be minimized. In recent years, however, the requirement for longer spans has meant larger horizontal reactions and has made rigid frames less economical. Figure 2-9 (page 32) shows a recent long span rigid frame where the sides of a rock cut provide an economical way to handle the horizontal reactions. Abutment restrained girders are now frequently used for sites where a thin and pier-free mid-span is desired. These are essentially three-span bridges with the short end spans hidden in the abutments (Figure 6-3). Structures with delta-frame center supports and slanted legs are possible and can provide a structure with greater-than-usual memorability. Figure 4-19 (page 110) shows that delta frames can work for highway overcrossings.

FIGURE 6-3 *The thin girder permitted by this abutment-restrained design opens up the view through the structure.*

Arch structures are rarely used for highway overcrossings because the required spans are too short for an arch to be economical, and the vertical clearance is usually not sufficient to offer a reasonable span-to-rise ratio for a deck arch. A through-arch structure may be a good solution where long spans and a very thin floor system are required, as in a single-point urban interchange (SPUI) (Figure 6-3). Figure 1-13 shows the use of a through arch to solve a restricted clearance problem in Houston.

Pier Placement and Abutment Placement

These bridges will be seen all at once, and the placement of the abutment strongly influences the placement of the piers, and vice-versa, therefore it is best to consider pier and abutment placement at the same time.

The pattern of the undercrossing roadways sets the possibilities for pier placement. The most common situations are:

- No pier at all (one-span structure)

- Median of an undercrossing dual roadway (two-span structure)

- Both outer edges of the undercrossing roadway (three-span structure)

- Median of an undercrossing dual roadway and both outer edges of the undercrossing roadways (four-span structure)

All of these solutions establish a basic symmetry for the structure, which must be respected in all further decisions (Figure 6-4). (The symmetry will be weakened if the structure is highly skewed, on a continuous vertical grade, or if the dual roadways are of substantially different widths.) Each solution also has implications for the apparent openness of the structure that must be considered. Compare the three bridges in Figures 6-5, 6-6, and 6-7, which are basically similar bridges except for the positions of the piers and abutments. Ramp splits, weaving and merging areas, and traffic control devices are important to safety and should not be blocked from view. A scenic focal point (a distant building or mountain), represents a rare opportunity to create an event on the journey, which should be recognized whenever possible. For an excellent example, Figure 6-15.1, page 205).

FIGURE 6-4 *Two of the possibilities for pier placement, showing the basic symmetry that results.*

FIGURE 6-5 *The shoulder—edge abutments interrupt the space of the highway and make the driver feel crowded.*

With both ends of the bridge in view at the same time, and with so few other elements in the structure, the placement and resulting proportions of the abutments will be critical to the visual impression (Figure 6-8). The key guidelines were given in Chapter Four. The most important is to relate abutment height to clearance at the roadway edge and to girder depth.

Skewed bridges have characteristics of their own that require special consideration. There are two guidelines that ease the problem. The first is to select superstructure types that allow large overhangs and narrow piers, such as bridges with an integral pier cap and/or box girders. The second is to move abutments to the tops of embankments and place them at right angles to the overcrossing roadway; see Figure 4-39, (page 123).

Superstructure Shape, Parapets, and Railings

Highway overcrossings have so few elements that the fascia girder and parapet become major factors in forming the visual impression. Small differences in proportion and detail can be critical, so it is important to design the parapet-girder combination together. The goal is to emphasize apparent thinness and horizontality to fit in with the dominant horizontal dimension of the typical highway environment. Application of the guidelines from Chapter Four will result in a girder-parapet combination with a pronounced horizontality and a thin appearance.

FIGURE 6-6 *Substituting shoulder piers for the tall abutments opens up the space somewhat, but still leaves the driver feeling crowded.*

FIGURE 6-7 *Eliminating the shoulder piers completely opens the view and eliminates the hazard.*

FIGURE 6-8 *The right abutment is much larger than the left, which emphasizes the effect of the vertical grade and gives the bridge an unbalanced appearance.*

Highway overcrossings are seen at one glance, meaning that any abrupt changes in girder depth will be jarringly obvious. Constant or smoothly varying depth will make the structure more unified. It will also make the structure appear to be longer, and therefore thinner. Structural continuity is a big help here, as it keeps everything visually continuous as well (Figure 6-9).

Haunched girders express the forces in the structure, and will provide an important point of interest that reinforces the basic symmetry of the structure. If a haunch is not feasible, the structure will depend primarily on the proportions of the girder for its interest, particularly its relative thinness, as compared to the other features of the structure. The series of Wisconsin overpasses shown in Figures 6-10.1, 10.2, and 6-10.3 show how a haunch gives an otherwise ordinary overpass some visual distinction, opens up the view, and improves safety by eliminating the shoulder piers.

Shaping rigid frames and delta frames is difficult at the relatively low heights of most highway overcrossings, hence requires exploration. Much depends on the relative lengths, vertical clearance, and shape of the side slopes (Figure 6-12). Written guidelines can be misleading. Better to make sketches of multiple shapes, as seen from the important viewpoints, then choose the one that seems most graceful.

Pier Shape

The primary visual goal for highway overcrossings is an appearance of a clear and safe passage through the structure, with a clear view to what lies beyond. The pier

FIGURE 6-9 *A well-designed highway overpass. The parapet overhang creates a horizontal shadow line that reduces the apparent thickness of the parapet girder combination; the appearance would be improved if the top edge of the pier were chamfered so as not to interrupt the line of the girder's bottom flange.*

FIGURE 6-10.1 *The initial type for these three overpasses was this four-span structure.*

FIGURE 6-10.2 *Perhaps in an effort to deal with a slightly longer span, this haunch was added. However, the haunch seems too small and provides little visual benefit.*

FIGURE 6-10.3 *The realization came that the haunch provided the possibility of eliminating the shoulder piers, which greatly improved the safety and appearance of the structure at little or no additional cost.*

FIGURE 6-11 *This structure shows that haunches can be done with precast concrete as well. The piers are unfortunately heavy for this graceful superstructure.*

FIGURE 6-12 *This frame's legs are too short.*

FIGURE 6-13 *This pier cap, and particularly its end elevation, adds a complication and a distraction; the pier would have been improved if the first column had been moved slightly to the right and the sloped face extended up to the girder, thereby eliminating the pier cap end.*

FIGURE 6-14 *A skew makes the pier more visible than the usual highway overcrossing pier.*

should not distract or detract from this goal, but should do its job of support as simply as possible. The most likely distraction is the end elevation of the pier cap (Figure 6-13). Chapter Five provides guidelines for pier designs that address this issue and others. Because these bridges have so few visual elements, the shape of each element and the proportions of each to the other take on great importance and must be carefully designed.

The side view of the far-side pier of a skewed structure (and a long-span multi-span structure) is more visible than the side of the center pier of a two-span bridge. Thus, pier design becomes even more important on skewed and multi-span bridges (Figure 6-14).

Abutment Shape

Because the abutment and parapet are such major elements, and because both ends of the bridge can usually be seen at once, the decision about how to shape the abutment has a major impact on the appearance of highway overcrossings. Chapter 5

FIGURE 6-15.1 *Traveling west on I-70 from Denver the driver's first view of the Rocky Mountains is framed by the Gunnison Road bridge. Note how exposing the girder seat on the abutment appears to stretch the girder and make it seem thinner.*

FIGURE 6-15.2 *The abutments of a similar structure in Wilsonville, Oregon, slice off the girder and make it seem both shorter and thicker. The sign exacerbates the problem by interrupting the lines of the bridge, making it seem even shorter.*

shows several good examples. The slope of the abutment face can be a major influence on the appearance of the structure. Bearings are a major structural element, and if left exposed will make the structure look longer and thinner (Figures 6-15.1 and 6-15.2). The end of the parapet profile and pedestrian screen at the abutment are also important considerations. The guidelines in Chapter Five should be followed. For most bridges, simplicity should be the goal.

Color

The opportunities for making an impact with color on highway overcrossings depend on the features seen in elevation: the fascia girder, parapet face, railings, pedestrian screen, abutment faces, and end elevations of any piers. A bold color or pattern can make a major impact, producing a more memorable structure than the size of the structure would normally allow. And, because highway overcrossings are often seen in quick succession, designers can use color, pattern, and/or texture to provide a theme that ties the group together.

With conventional steel girders, the necessity of painting the girder makes a color choice necessary and offers a "free" opportunity to add interest. With A588 steel and concrete surfaces, the application of color can be avoided, but the combination of light gray and dark brown may not suit the surroundings, the desires of the community, or the designer's intention. If most of the structural surfaces will not be coated, smaller accent areas of color are a good way to add interest or to complement coated surfaces. A "racing stripe," inset in the parapet and coated, is one frequently used technique. The bright-red stripe on the bridge in Figure 5-46 (page 173) makes it a memorable element, while reducing the apparent depth of the parapet. Insetting materials such as glazed tile is another possibility. The use of color on a railing or pedestrian screen is a good way to add interest and make these elements something more than utilitarian extras.

By dividing a girder horizontally into several areas of different but compatible colors, the girder can be made to seem thinner. This works best when a horizontal stiffener is present to provide a logical division line (Figure 6-76, page 270).

Because the entire bridge is seen at a glance, it is important that all color decisions be considered at the same time and coordinated. Figure 1-23.1 (page 22) shows a pedestrian screen of one color on a bridge painted with a different color scheme. The orange of the screen is the complementary color of the blue of the girder. Complementary colors will always draw attention and create visual tension. That is acceptable if anticipated and desired, but is difficult to pull off in a single structure. A color more closely related to the girder color (a lighter shade of blue) or a step away from the complementary color, (red rather than orange) would have worked better for the fence.

Surface Patterns, Texture, Ornamentation, and Details

When the undercrossing roadway is a high-speed roadway, the effectiveness of patterns is limited unless the units are large. Any detail less than 4 inches in size will be missed. The faces of the parapet and the higher abutments offer opportunities to add interest through the addition of patterns. These can be abstract or representational and can be repetitive or singular. Chapter Five illustrates several examples.

Any pattern must be kept within, and consistent with, the geometry of the structural feature of which it is a part (Figure 5-49, page 178). If a pattern bears no relation to structural features, it will divide the structure into visually separate but physically connected objects. Because highway overcrossings are seen at one glance, the dichotomy will be noticed and will seem jarring.

Traditional masonry materials are often considered for abutment faces and piers, either to recall nearby historic buildings or to establish consistency with associated natural features. These can be satisfying features as long as they are used in ways consistent with their inherent structural capacities and historical uses. That means restricting their use to pier or wall facings. Using such materials as a girder facing puts heavy masonry in the air with no visible means of support. People don't understand how it can work and become uncomfortable.

For structures over high-speed roadways, larger masonry units are necessary to maintain some sense of texture and differentiation (see Figure 3-13, page 77). Smaller units blur into a single surface color. Structures over streets and arterial roadways offer more scope for smaller-scaled textures and materials. If pedestrians are present, the use of interesting materials at fingertip range will enrich and enhance their experience and will be appreciated.

Signing, Lighting, and Landscaping

Highway overcrossings are often called on to serve double duty as sign bridges. The visual result is almost always unfortunate for all the reasons outlined in Chapter Five. It is a particular problem for highway overcrossings because these bridges can be seen at a glance and the presence of the sign is immediately obvious. Also, the sign covers a large portion of the total structure, and is usually placed without regard to the basic symmetry of the bridge or the size of the sign relative to structural features.

The best solution is to find another location for the sign—on an elegant sign structure separate from the highway overcrossing. Rarely are the rules for sign placement so inelastic that the highway overcrossing offers the only possible location (Figure 6-16). Even 300 feet upstream is enough of a separation to provide an effective

FIGURE 6-16 *There are many other options for the location of this sign away from the structure. These three bridges mix two different girder shapes and two different pier shapes. The group would have been improved if the deisgner had settled on one.*

view of the bridge. The second-best solution is to design the sign panel to fit the features of the bridge as described in Chapter Five. The effort is particularly worthwhile for highway overcrossings. If a sign will be present, it will have a significant influence on the view of the structure, hence should be considered in any color choices that are made.

Lighting can also have a major effect on the appearance of a highway overcrossing. Typical highway lighting poles are twice as high as typical overcrossings and represent sizable physical features when compared with the bridge itself. If placed without a clear relationship to the structural features of the bridge, they can seem very much out of place. The rules outlined in Chapter Five are particularly important here. Because highway overcrossings are relatively short, usually it should be possible to adjust the spacing of fixtures for an overcrossing roadway so that a pole location coincides with a central pier location. Other poles will then have a symmetrical relationship consistent with the basic symmetry of the bridge.

When the overcrossing roadway is a street with pedestrian sidewalks, there is sometimes a desire to provide more numerous fixtures at a lower height, often with ornamental features. The supports for these fixtures need to be designed together with the railing, and any ornamental features of the lighting must be consistent with the ornamental features of the railing. It is even more important that such lighting be lined up with basic structural features, because they will establish a dominant visual spacing that will make anything not in alignment with them seem out of place. Such fixtures can indeed establish a memorable appearance for the structure both day and night.

The lighting of the undercrossing roadway is generally not a concern in most highway situations unless the overcrossing roadway is very wide. But when pedestrians are present, the lighting of the undercrossing roadway is very important. The basic rule is to arrange the fixtures in some consistent relationship to the major structural features, such as one fixture for every third girder. Lighting the adjoining surfaces of the structure, such as an abutment wall or the underside of a box girder, can make the area seem more attractive and secure than lighting the sidewalk only.

If the structure is serving as a gateway, it may be useful to light the structure itself. One approach is to light the elevation surfaces: the fascia girder, parapet, and abutment faces. Another is to light the underside of the superstructure and the abutment walls. That produces an arch of light and can be a very effective welcome. Doing both together may seem like too much light.

Landscaping can significantly enhance the appearance of highway overcrossings. The overcrossings are typically part of largely man-made environments of graded slopes that will be replanted in any case. The structure is not so large as to overwhelm any landscaping that may be attempted. Since the structure can be seen all at once, the role of landscaping in the overall scene can be understood and appreciated. Planting can be seen as a technique to "stretch" the structure and make it appear longer, and therefore thinner. Planting can also be used on the undercrossing roadway to lead the driver's eye through the structure and to emphasize the continuity of the scene (see Figures 2-55, page 57, and 5-69, page 193).

CASE STUDY

Maryland Route 103 over Maryland Route 100, Columbia, Maryland

Maryland Route 100 is a six-lane freeway connecting Howard and Anne Arundel Counties, suburbs of both Baltimore and Washington, D.C. Maryland Route 103 is an existing two-lane arterial that had to be re-connected across the freeway. It includes 10-foot shoulders, but no sidewalks. The area is characterized by rolling, wooded terrain covered by town-house and single-family residential areas, low-rise office and industrial parks, and shopping centers. Much of Maryland Route 100 is in cuts of varying depths. At the bridge site, Maryland Route 103 is essentially at the elevation of the surrounding topography; Maryland Route 100 is 25 feet lower. The sides of the cuts slope at a ratio of 2 feet horizontally to 1 foot vertically.

The background against which the bridge is seen consists mostly of the vegetated slopes of the highway cut, which show significant regrowth of trees and shrubs. The tops of nearby office buildings or houses can occasionally be seen above the tops of the vegetation. The predominant colors are the multiple shades of green typical of mid-Atlantic vegetation and the occasional light beige and white of distant buildings. The primary viewpoints for the bridge lie along the driving lanes of Maryland Route 100. There are no nearby buildings or other locations with more than an incidental view of the bridge. The nearby uses are all auto-oriented, and most are at some distance from the bridge, resulting in little pedestrian and bicycle traffic.

The Maryland Route 103 bridge is one of a number of similar crossings of Maryland Route 100 along its 25-mile length. It has no special symbolic importance to its neighborhood or as a milestone along Route 100. Indeed, the Maryland State Highway Administration's overall approach to Route 100 was to blend it into its surrounding area by, for example, providing wide recovery areas and gradual, vegetated side slopes. The boundaries of the bridge are set by the tops of the cut slopes, which cut off the view of Maryland Route 103 as it proceeds away from the bridge.

The design intention/vision focused on the appearance that the bridge presented to travelers on Route 100. The goal was to keep the view through the bridge as open as possible, so that the space of the highway appeared to continue right through the bridge (Figure 6-17). Because this is one of a number of similar bridges, a design theme was created based on the use of steel girders, and includes similar, simple shapes and materials for the piers, abutments, and parapets. The criterion was to have a consistent appearance for all of the bridges, so as not to call attention to any one.

FIGURE 6-17 *MD 103 over MD 100, Columbia, Maryland, allows the space of the highway to flow right through the structure.*

FIGURE 6-18 *The curved edges of the pier eliminate the pier cap and appear to continue the lines of the haunch when seen at an oblique angle.*

The geometry in this case was set by the alignments of Maryland Route 103 and Maryland 100 and did not present either with any notable problems or opportunities. The choice of steel welded-plate girders for the whole route was based on the fact that the bridges all have similar span lengths and that none of the bridges need to accommodate pedestrian traffic below, so that the appearance of the underside was not a concern. At least one pier was necessary to avoid an unreasonably long span, with the center of the median being the logical location. In view of the design intention to keep the view as open as possible, as well as the safety problems associated with shoulder piers, none were considered. Similarly, the abutments were placed as high as possible on the side slopes to allow the slopes to appear as continuous surfaces flowing right through the bridge. The short abutment wall also makes inspection of the abutment easier. The girder was haunched to make it appear thinner and to show the flow of forces. An overhang equal to one-half of the girder depth at mid-span was used to create a shadow line that further minimized the depth of the girder and reduced the longitudinal width of the pier. There is a 5-foot-tall pedestrian fence on both parapets.

The piers were designed as a series of Vs with curved sides (Figure 6-18). This design eliminated the pier caps as a visual element and gave the piers a shape that appears to continue the curve of the girder haunch when the bridge is seen at an oblique angle. A screen wall was provided at the abutments to hide the bearings. The edge of this wall is angled inward to reflect the curves of the piers. The girders are

FIGURE 6-19 *U.S. 1 over MD 100, Columbia, Maryland. Even though most of the basic elements are similar in shape to the MD 103 bridge, the effect is quite different. The high abutments narrow the space of the highway, and the straight girder seems ordinary. The signs interrupt the lines of the bridge and further degrade its appearance.*

painted brown, and the concrete is uncoated. Form liners are used to create texture on recessed piers and abutment panels to make the recesses seem deeper. The face of the parapet is divided into two horizontal facets, one angled upward and one angled toward the ground. The upper facet catches more light and is relatively bright, while the lower one reflects only ground reflections and seems relatively dark. The two horizontal bands of different brightness make the parapet seem thinner and add interest to the structure. There are no signs or lights on the structure. The landscaping associated with the structure is only that necessary to reestablish the natural vegetation of the area.

All of the elements used in the design of this bridge are within the range of the standard designs used by Maryland. The agency feels that the aesthetic aspects of the bridge had little or no effect on its cost.

The structure is very successful in meeting its design intention of providing open views in which the space of the highway continues right through the structure. While not calling undue attention to itself, the bridge appears to be a thin and graceful structure that shows how it works. To see how important the basic lines of the structure can be to its visual success, one only has to compare it to an earlier version on the same route, US 1 over Maryland Route 100 (Figure 6-19), where different choices about abutment placement and girder shape resulted in a very different effect, even though similar elements were used for the piers, abutments, and parapet of the bridge.

The dark brown of the fascia girders (similar to the color of A588 steel) does not fit well with the variegated greens of the background and gives the bridge a heavy, somber look. The brown would look more in place in the desert environment of the Southwest. Opening up the implied voids in the piers, as in the Brainerd Bridge, Figure 2-7, would have given the bridges a lighter, more transparent appearance. The chain-link fence seems unnecessary given the lack of pedestrian/bicycle usage.

The Maryland Route 103 bridge was designed by the bridge design staff of the Maryland State Highway Administration. Paul Matys was the project manager.

CASE STUDY

The I-65 Gateway Arch Carrying I-65 over a Single-Point Urban Interchange, Columbus, Indiana

Highway-over-highway bridges at single-point urban interchanges (SPUI) can be a real challenge to the bridge designer. When the freeway is above the arterial street, long spans (200 feet-plus) are often required to cross the street and its curving left turn ramps below. The usual response to this is a single-span girder, which gets very deep and appears massive. The greater vertical distance from lower roadway to the bridge deck requires lengthened ramps and, often, lengthened retaining walls. The abutments are usually placed close to the roadway to minimize the span. The high abutments restrict sight distance for turning drivers and, combined with the massive girder, give the bridge a "hole in the wall" look (Figure 6-20). The bridge can be made a three-span structure, creating more sight distance and allowing a thinner girder. The cost of the lengthened superstructure is often offset by the savings in abutment and retaining wall costs (Figure 6-21). Another approach is to use structures with thin floor systems, where the additional cost can be offset by savings in ramp length, right-of-way and retaining wall costs. This case study describes one such example.

Different problems arise when the main roadway is below the structure. Then the plan shape of the structure is complicated by the need to support the curving left-turn lanes. If straight steel or precast concrete I-girders are used, complicated and unattractive framing plans result, with variable overhangs and shadow lines. This is the place for curved girders or concrete box structures, as shown in the Phoenix, Arizona, case study, next.

The genesis of the I-65 bridge (Figure 6-22) was the need to rebuild an existing diamond interchange into a single-point urban interchange. I-65 was an existing freeway carrying four lanes of traffic. There are several other bridges nearby on I-65, including a sizable river crossing. The undercrossing roadway required visual and physical clearance, both for its four through-lanes and for the opposing left-turning movements. The arterial is one of the main entrances from the Interstate system to Columbus, Indiana. There is a minor amount of pedestrian and bicycle traffic along the arterial street.

The area surrounding the bridge is flat, partially wooded, and contains a mix of residential, commercial, and office uses typical of a suburban commercial arterial. Because of its location at one of the major gateways into Columbus, there was a desire to have a landmark at the location, both to announce Columbus to travelers on I-65 and to provide a focal point for the variety of activities along the arterial street. The raised embankment carrying I-65 is visible for some distance from the surrounding area, as is the arterial street itself (Figure 6-23). The bridge appears as an extension of I-65. There are no clear boundaries.

FIGURE 6-20 *The deep girder and high abutments give this structure, built for a single point urban interchange (SPUI), a "hole in the wall" appearance.*

FIGURE 6-21 *The open sidespans improve the sight distance and the overall appearance of this Denver, Colorado SPUI bridge.*

The design intention for the structure focused first on providing the long span required, without creating a need for extensive reconstruction of I-65. The second goal requested by the town was to develop a bridge that could serve as a landmark for Columbus.

Because of the long span required to bridge the single-point urban interchange, a conventional girder bridge would have required a deep girder. Creating sufficient vertical clearance for the girder would have required raising I-65 as much as 3 feet, affecting I-65 for a half-mile in either direction and requiring reconstruction of the other nearby bridges. Thus, attention was turned to structural types with a shallow floor system that would leave the profile of I-65 untouched. Through structures of various types were considered, but the basket-handle arch seemed to offer the best combination of simplicity and cost. The arch was placed in the median, where its two halves brace each other, so that clearance problems of overhead bracing are avoided. The curve of the arch was carefully chosen to keep the resultant force at each point within the arch and minimize moments, which resulted in a compound curve.

The roadways are carried by floor beams cantilevering in both directions, creating the shallow floor system required. The underside is kept completely free of piers. The abutments are set high on the side slope and integrated with the arch footings. The open underside allows free sight

FIGURE 6-22 *The I-65 Gateway Arch, Columbus, Indiana. The selection of the structural type was driven by the need to provide a long span to accommodate a new SPUI without raising the profile of I-65. The arch provided a 200-foot span with a floor system depth of 2 feet.*

FIGURE 6-23 *This arch creates a landmark along I-65 that signifies one of the main entrances to Columbus, Indiana (see color image).*

distance in all directions for both through and turning traffic. The bracing of the arches and the detailing of the suspenders was kept as open and simple as possible, so that the arch is easy to see through (Figure 6-23). The arch provides the landmark element that the town was seeking.

A bright red color was chosen for the arch to signify its desired landmark status. No other texture or ornamentation was necessary for it to achieve that status, and none was added. The bridge was kept free of lighting and signs to avoid compromising the lines of the arch. The infield areas of the ramp were landscaped to suit a gateway interchange.

The bridge cost $180 per square foot in 1998, which was actually about the same as the other alternatives available at the site when total costs were considered, including necessary changes to I-65. Construction time was about the same as for a conventional bridge. The arch consists of steel pipe that was induction-bent to the required curve. No American Institute of Steel Construction (AISC)-certified fabrication shops do induction bending, so that was done in an American Society of Mechanical Engineers (ASME)-certified pipe and boiler fabrication plant. In these shops, the welding and material control standards are actually more stringent than in AISC shops, so the quality of workmanship was not a concern. Since the arch is always in compression, it is not fracture-critical. The arch rib is sealed during fabrication like any other box or tubular member, and no further maintenance is expected. Because there is no bracing, there is a minimal area to paint, and no corners or ledges to catch dirt or pigeon dung. The cable suspenders are galvanized structural strand with conventional fittings, which have been in use on arch and suspension bridges for decades. The floor system is designed so that one suspender can be removed for repair or replacement without disturbing the remainder of the bridge. The deck can also be replaced without disturbing the remainder of the bridge. No unusual maintenance problems have been experienced or are expected.

The bridge has been a success in meeting its design intention of bridging a single-point urban interchange without requiring the reconstruction of I-65. In the course of solving this difficult engineering problem, it also created a landmark for the city of Columbus.

The bridge was designed by J. Muller International, now part of Earth Tech, for the Indiana Department of Transportation. Dan Burroughs was the bridge design manager.

CASE STUDY

Cactus Road over the Pima Freeway, Phoenix, Arizona

SPUI interchanges with the arterial street on top require a bridge that varies considerably in width to accommodate the left-turn lanes (Figure 6-24). They also often require retaining walls to minimize the right-of-way required. The challenge is to smoothly connect all of these features so that the project appears as a single entity, not a collection of disparate parts.

The Pima Freeway extends through developed and developing suburbs of Phoenix and Scottsdale, Arizona. The freeway is essentially straight and is depressed below grade in the section containing the interchange. Parallel frontage roads occupy the surface along the freeway's edges. The surrounding area consists of low residential and commercial buildings. For the most part, only the tops of the buildings are visible from the freeway. There are frequent retaining walls between the frontage roads and the freeway, but these are set at the edge of the frontage roads, allowing for graded slopes along the edge of the freeway. This gives travelers on the freeway more space, making the walls seem less confining. Cactus

Road is one of a number of interchanges with similar arterials along the freeway and there was no desire to make it a particular landmark or gateway.

In order to unify the experience of the freeway and make the trip more attractive to users, a theme was developed for the highway. All of the elements of the highway were designed together, using continuous shapes and similar colors and patterns for bridges, walls, slopes, and other features so that they all clearly relate to each other. The theme includes the use of southwestern patterns and colors on bridges, walls, and the desert landscaping, which uses patterns in colored gravel for the side slopes with occasional desert plants. Reinforced concrete box girders are used most frequently on the freeway, primarily because their minimal structural depth minimizes ramp length and retaining wall height and because they can smoothly adapt to changing widths and branching (Figure 6-25). They also allow for a sizable deck overhang, which makes the structures seem thinner, and narrows the width of the substructure. The design intention/vision for the Cactus Road overpass was to fit into this theme as seamlessly as possible.

The Cactus Road overpass bridges a divided highway of six lanes with shoulders. It is a two-span structure with the piers in the median and high abutments set back 30 feet from the edge of the traveled way to avoid the need for guardrail. The spans of the bridge are 90 feet. The bridge carries a four-lane arterial street and must also provide for four two-lane left turn ramps. For this reason the plan of the bridge is hourglass-shaped and varies in width from 130 feet at the piers to 199 feet at the abutments. The geometry of the left turn ramps requires the edge of the bridge to be continuously curved. The parapet of the bridge plus the parapets of the adjoining retaining walls actually make a 180-degree curve. The bridge is a reinforced concrete box girder 5 feet 6 inches deep with eight webs. The outer webs slope outward at 1 horizontal to 3 vertical and follow the curve of the parapet with a constant overhang of 3 feet 3 inches.

The geometry of the bridge was determined by the profile of the freeway and the intersecting arterial frontage roads, leaving no room for adjustments and no real reason for them. The only reasonable pier location was the median. The abutment locations were set by the need to end the bridge before the widening required by the left-turn lanes became too great. The abutment wing walls were aligned to extend the curve of the bridge parapet and blend smoothly into the retaining walls and safety barriers along the frontage roads, visually integrating the structure into the overall design of the freeway. The superstructure shape was set to match the other structures on the route, particularly the consistent deck overhang. The parapet profile was carried from the bridge along the tops of both retaining walls, further reinforcing the integration of the bridge with the rest of the project.

FIGURE 6-24 *Cactus Road over the Pima Freeway, Phoenix, Arizona. The curved-edge barrier for the left-turn lanes blends smoothly into the bridge parapet, which in turn blends smoothly into the retaining wall parapet, creating one continuous line.*

FIGURE 6-25 *The fascia of the concrete box girder follows the curve of the parapet, making the whole structure seamless. The pattern on the abutment wall fits within the lines of the structure and works well. The pattern on the pier, in contrast, is too big for the pier, and is wasted in any case because it is on the side of the pier, which drivers don't see.*

The pier is faceted, being an elongated octagon in plan, which makes it appear thinner to oncoming drivers. The poured-in-place retaining walls are made up of rectangular sections that lend themselves to the color and pattern scheme used throughout this section of the freeway. A desert lizard pattern is cast into a panel on each side of the bridge and into the pier side. The wall patterns work well: They are large enough to be understood and in a location where they can be seen by approaching travelers. However, the lizard pattern on the side of the pier is too large for the pier and breaks up the lines of the pier when seen by approaching travelers. Fortunately, being on the side of the pier, it cannot be easily seen or understood by approaching travelers, so it has a minimal effect on the overall appearance of the bridge.

The Cactus Road Bridge meets its design intention of blending seamlessly into the theme of the Pima Freeway while efficiently and elegantly handling its structural role. The bridge was designed by HDR Engineering, Inc., for the Arizona Department of Transportation.

VALLEYS AND HIGHWAY CUTS

This section focuses on sites where the height of the structure is large relative to its length, a situation that frequently occurs in ravines, valleys, and highway cuts.

Developing a Design Intention/Vision

Physical Requirements

By definition, sites for this type of structure are physically demanding: high vertical clearances, steep and/or long slopes, often with environmentally sensitive features, and waterways. Determining site requirements involves doing a complete review of the available environmental and geotechnical reports and conducting interviews with representatives of other agencies with interests in the area. Avoidance of roadways, wetlands, and waterways may restrict pier location. Utilities may be a concern, both as to pier location and as a requirement that the structure carry one or more utilities. There may be restrictions on the disposition of drainage from the structure that will affect the design.

Visual Environment

The visual environment will be dominated by the nature of the slopes adjoining the structure. If they are natural, they should be viewed and photographed. Tree cover, the extent and nature of rock outcroppings, and the steepness and shape of the slopes will all have a bearing on later decisions. If the slopes will be created by a highway project, their nature can be determined from the highway plans and replanting practices. It may be possible to influence the slope design to create a shelf that would become the logical point for the placement of an arch or rigid frame's abutment. A personal visit is necessary to become familiar with the larger area and to determine from which areas and viewpoints the new bridge will be visible.

Nearby and Associated Uses

The bridge itself will be insulated to some degree from immediately nearby uses by the adjoining slopes. However, because of its size, it will have a dominating effect

FIGURE 6-26.2 *I-470, Wheeling, West Virginia. An attractive structure will create an event along the freeway that people will use to measure their journey. A steep valley or highway cut is a natural location for an arch or rigid frame. The simplified bracing is clearly visible, which would not be evident on an elevation drawing.*

FIGURE 6-26.1 *By using long spans with a single column at each pier line, this Tennessee structure keeps diagonal views through the structure open.*

on any nearby use area from which it is visible. A frequent situation is a bridge over a natural area, such as a park. The park users will see the bridge as a dominant landmark and it will influence all future enjoyment of the park. A high bridge near a town will become a dominant landmark for the town.

High bridges over highways, because of their position and size, will be a memorable event, a milepost by which people will measure their journeys.

Symbolic Functions

Because of their size and prominence, these bridges will often assume symbolic functions that should be anticipated. A high bridge crossing a highway near a town will be seen as the gateway to that town. A high bridge that is visible from the center of a town can become, for better or worse, the defining landmark for that town. A bridge over an important natural area will become a symbol of respect for that area if sensitively done or, if not, a symbol of carelessness.

Boundaries

The slopes that define these bridges set the boundaries. The size of the bridge usually dwarfs any associated highway elements, so the bridges can be effectively considered on their own.

Guidelines for the Design Intention/Vision

The first consideration is the position of likely observers. In the event the bridge is also spanning a roadway, particularly a bridge spanning a major highway cut, many of the observers will be in cars on the underpassing roadway. However, the extended length and much greater height make this a much different situation from the ordinary highway overpass. The greater size ensures that it will be seen from a greater distance than a typical highway overpass, so it will be in the visual field for a longer time and thus make more of an impact. The greater height and longer span also mean that the underside of the bridge and the sides of the piers will be more visible and more important in creating the impression. If the area under the bridge will be used by pedestrians, the appearance of the underside will be particularly important. The greater size of the bridge means that the abutments will be a smaller proportion of the bridge and less important to the overall impression.

For this type of bridge there will be a number of important observer locations, in addition to any underpassing roadway. Some of these viewpoints may be a mile or more away. While the designer can't cover them all, he or she at least should be aware of the most important, and consider these viewpoints when decisions are made. Any photos, drawings, or sketches should be taken from at least two or three of the most important viewpoints.

From any viewpoint, the shapes of the major structural features of the bridge are likely to determine the visual impression. The elements to be most concerned about, in order of importance, are the shape of the horizontal and vertical geometry, superstructure type and shape, pier placement and shape (or, to look at it another way, the number and shape of the openings), and the shapes of the piers. Parapets, railings, and abutments will be a smaller portion of the total structure and have less influence on the aesthetic impression, though the normal criteria concerning simplicity and consistency with the rest of the structure still apply. When designing a bridge over a valley or highway cut:

- Develop an overall shape and color of the major structural members that is memorable and that complements their surroundings.
- Provide generous vistas through the structure.
- Place piers in visually logical positions within the site.
- Aim for apparent thinness in structural members.
- Use the shapes and details of secondary elements to reinforce the shapes of the main structural members.

Determinants of Appearance

Geometry

The bridge will look best from surrounding areas if it is straight or composed of a few long, continuous curves (Figure 6-27.1). Any consideration of geometry also needs to take into account the driver's view along the bridge deck. This will also generally be best if the curves are long, continuous, and generous (Figure 6-27.2). In particular, placing two short vertical curves on either end of a bridge, with a tangent on the bridge in between, will produce a very uncomfortable view for the driver. This is often done to avoid a sag vertical curve on the bridge. Sag vertical curves on bridges are visually acceptable if they are long enough.

The creation of a drainage sump on a bridge is undesirable for maintenance reasons. Careful coordination with the highway designer will usually make it possible to

FIGURE 6-27.1 *A long, continuous curve creates an attractive geometry. This structure also demonstrates a simple but effective solution to a family of tall piers. A large deck overhang pays visual dividends here because it keeps the piers narrow and the girder in shadow.*

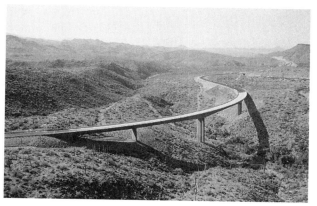

FIGURE 6-27.2 *The long spans and thin piers for the box girder supporting the Lower Screwtail Bridge in Sunflower, Arizona, allow the desert landscape to flow through the structure. (See also color insert.)*

move the actual low point off the bridge even though the balance of the curve extends well onto or across the structure. Long structures of this type may contain superelevation transitions, which should be checked for their visual effect on appearance of the parapet.

Superstructure Type

Valleys and highway cuts present most of the opportunities for a deck arch or rigid frame. The surroundings "contain" the visual thrust of the arch and frame legs, and there is enough vertical height to allow frame legs of a reasonable length. The choice of structural type should depend first on the structural economics of the required bridge. That said, arch and frame forms offer a visual dynamism that may weigh in the balance at particularly prominent or important locations (Figures 6-26.2, page 217).

Because the overall height of bridges over valleys or ravines is so great, the depth of the superstructure is generally not critical, and girders and even deck trusses are also reasonable solutions. Pier placement and girder or truss shape then become the key elements in determining the visual impact.

The appearance of the underside may be important if the bridge is seen from below. The usual confusion of wind bracing and diaphragm bracing required by I-girders is a potential distraction because of the number of members, the number of different angles at which they are installed, and the visual competition they offer to the main lines of the structure (not to mention the additional roosts for birds). Concrete or steel box structures, where the underside is completely enclosed, solve this problem. Girder-type bridges can be satisfactory as long as the details of diaphragms and wind bracing are kept simple and consistent along the bridge.

The choice between a steel and concrete superstructure is a visual, as well as a structural, issue because the color of the structural members, when seen against their background, is so important in creating the visual impression of this type of bridge. Ordinary steel can be painted to suit the situation. If A588 steel is to be left exposed, the dark brown color will strongly affect the impression the bridge makes. The same is true for the light gray color of concrete. The effect that these colors will have in forming the impression of the bridge from the major viewpoints should be considered in making the choice of structural material.

The shape of the valley may require large variations in span lengths and even changes to a different superstructure type. It is important to maintain continuity of structural form, material, and secondary features such as parapets and railings.

Pier Placement

For all valley and highway cut bridges, locating piers and footings requires consideration of the overall shape of the valley or cut, of the clearance requirements and obstacles, and of the height of the bridge. Diagonal views from underside uses and nearby communities are usually important, and several piers are often visible from a given location. In order to prevent the view from degenerating into a wall or a forest of columns, the characteristics of groups of piers need to be considered. The guidelines in Chapter Four apply.

Abutment Placement

Abutment location is generally determined by the shape of the side slopes. In this type of bridge, the abutments are a small proportion of the total structure and have relatively little influence on the overall appearance. Best to keep them small (abutment height equals adjoining girder depth, more or less) and simple.

Superstructure Shape, Parapets, and Railings

The choices made about the superstructure shape will also have a dominant effect on the final impression made by the structure. Arch and rigid frame structures should be carefully shaped, using the guidelines in Chapter Four. On girder bridges, the spans will often vary in length over the bridge, making it necessary to change girder depth, provide haunches, or do both in a single bridge. It is important to keep structural depth constant or smoothly varied over the entire bridge (Figure 6-28).

FIGURE 6-28 *Upper Middle Road over 16 Mile Creek, Oakville, Ontario. The pier blends smoothly into the sloped sides of the box girder. The overlook balconies above the piers are too small; they are not consistent with the massive scale of the bridge.*

Details of shaping the parapet and abutments will not be as critical as for highway overpasses because these elements are a relatively smaller portion of the total bridge. Nevertheless, general rules of simplicity and continuity of materials still apply.

Pier Shape

Because of their size and the multiple locations from which they will be viewed, the piers will be major elements in forming the impression of the bridge. The piers will more than likely vary in height over the bridge length, therefore the key is to come up with a family of pier shapes that relate well to each other when viewed as a group. The goal is to portray a smooth flow of forces along the girders and from the girders into the piers and to the valley floor. The guidelines for tall piers in the "Pier Shape" section of Chapter Five will generally apply.

The pier cap is a prominent element in the oblique view, and it will interrupt the visual lines of the structure unless it is minimized. The length of the bridge causes this effect to be repetitive and obvious. This is particularly true if the end of the cap is near or at the same plane as the face of the parapet. Figure 6-29.1 shows a design that eliminates the pier cap end and maintains the flow of the girder line. The bridge in Figure 6-29.2 does almost as well by rounding the end of the pier cap. See the Pier Shape Section for other ideas for minimizing the pier cap end.

FIGURE 6-29.1 *The pier for the Lower Screwtail Bridge was split for flexibility to respond to thermal movements; but as a corollary benefit, the bridge becomes more transparent.*

Abutment Shape

Because of the small part they play in the appearance of these structures; abutments should be kept as small and simple as possible.

Color

Color is the final important element in the success of these bridges. Structures across dramatic valleys deserve colors that are as bold as the structures themselves. The goal should be to create a contrast to the background by using colors that enhance the background hues. A bright red-orange against a forested hillside is a classic combination. Figure 5-42.1 (page 167) shows an example of good color contrast. Sienna red against a brown western mountainside is another satisfying solution. Conversely, sometimes the position of a bridge in a national forest or similar sensitive area requires a more subdued approach. It may not be as desirable, for example, to have a bold color on a bridge that is seen predominantly as part of the tree canopy or that is visible from the town square. In this case, medium-hue, medium-value colors should be found that incorporate the colors of the area. By overlaying CAD-generated drawings on color photos many choices can be tested before a commitment is made to paint.

Bridges of uncoated concrete or unpainted A588 steel must "live" with the color of the structural material. Opportunities may exist to influence the impression made by that color by adding strongly colored smaller elements to the parapet face or girder, such as a painted stripe in the parapet face, or by adding color to a railing or light posts.

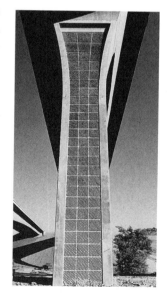

FIGURE 6-29.2 *Oak Canyon Bridge, San Diego, California. The horizontal lines of the grid pattern of the textured inset of this pier compete with the vertical lines of the pier.*

Surface Patterns, Texture, Ornamentation, and Details

Traditional brick and stone are generally not convincing when used as components of these bridges. These bridges are generally so much larger than familiar historic masonry structures that the use of such materials is not credible.

If a masonry material is absolutely necessary, large block ashlar stone is the one material with some visual and historic credibility. Natural stone is generally not economical, but it can be simulated with high-quality form liners with individually stained "stones," as described in Chapter Five.

The details of the girders offer possibilities to add interest in ways that are consistent with the structure. The arrangement of stiffeners on steel structures can be ornamental as well as structural. The details of bearing seats and pier connections can also be shaped to provide interest and make a positive contribution to the overall structure. These details can be further accentuated by the use of paint or inset materials of contrasting colors.

Signing, Lighting, and Landscaping

Sign structures are sometimes attached to the bridge for the benefit of the upper roadway. Such structures can be a distraction for these bridges. If it is absolutely necessary to place them on the bridge itself, they should be positioned in some logical relation to the overall bridge, such as at a pier, and smoothly integrated with the overall structure. The sign structure should seem like part of the original concept, not a bolted-on afterthought.

The presence of roadway lighting poles will provide a chain of lighted dots across the bridge, which will heavily influence the appearance of the bridge at night. Lighting poles should be spaced in some easily understood relationship to the main structural members—for example, at a spacing that is a constant fraction of the pier spacing. If the bridge is, or appears to be, symmetrical, the light poles should be symmetrically placed. Their supports should be smoothly integrated into the parapet/railing design.

Landscaping will be dominated by the natural conditions of the adjoining slopes, and is generally not a major consideration for this type of bridge. Any planting should be aimed at restoring natural conditions to areas disturbed by construction.

CASE STUDY

Truckee Valley Bridge, Truckee California

The Truckee Valley Bridge provides a bypass of the historic town of Truckee for travelers going south from Interstate 80 to resort areas in the Sierras (Figure 6-30.1). The bridge carries two lanes of California Route 267 over the valley of the Truckee River. The surrounding area is ponderosa pine forest typical of the High Sierras, with multiple rock outcrops and little vegetative understory. The predominant colors are the dark green of the pine foliage and multiple browns of the pine trunks, ground cover, and rock. Truckee attracts many tourists to its historic area. The bridge can be seen by travelers on I-80, from the historic center of Truckee, and by kayakers, hikers, and fishermen in the valley. The boundaries of the bridge are set by the pine forest.

The design intention/vision was to make a bridge that simply and gracefully crossed the valley without drawing attention to itself and with minimal interference to views up and down the valley. The geometry for the bridge was set by the alignment of California 267 and did not offer any particular problems or opportunities. A post-tensioned concrete box girder was chosen as the structural type because it offered long spans with relatively thin and simple piers. The piers were placed to provide an odd number of spans and to avoid the deepest part of the valley. The bridge is 465 meters long and 13.07 meters wide, carrying two 3.6-meter lanes with two 2.4-meter shoulders. There are no sidewalks. The bridge has seven spans; the end spans are 53 meters, and the remaining spans are 71 meters. Superstructure depth varies from 2.3 meters at midspan to 4.57 meters at the piers. The superstructure box girder has three webs; the outer webs are sloped at 1 horizontal to 3 vertical. The clear span at the top of the box is 3.18 meters. The overhangs are conventionally reinforced concrete and have a clear span of 2.9 meters. The bridge is 29 meters above the valley floor at its highest point.

FIGURE 6-30.1 *CA 267 over the Truckee Valley, Truckee, California, gracefully crosses the valley without drawing attention to itself.*

The abutments are small and are placed at the point at which the highway profile meets the existing grade. A haunched girder was chosen to minimize the thickness of the girder while permitting long spans, and to allow the bridge to respond to the flow of forces.

The piers are shaped to blend seamlessly into the haunched girders (Figure 6-30.2). The abutments have little impact on the appearance of the bridge and were kept small and simple. The concrete of the bridge was left uncoated. The

FIGURE 6-30.2 *The haunched concrete box girder flows smoothly into the piers (see color image).*

natural color of the local concrete has an affinity for the surroundings and the local sunlight, and the bridge seems to take on a rose beige color that blends in well with the background. With that in mind, and with the goal of making the structure blend in to the site, no further texture or ornamentation was needed and none was added. There are no inset pine trees in the piers or gold-rush pioneers on the parapet. There is no signing or lighting on the bridge, and the only landscaping is provided by the pine forest. The bridge cost $8.2 million in 2000. The cast-in-place post-tensioned concrete box girders are standard in California, and the owner believes that no significant premium was paid to achieve the aesthetic quality. Local elected officials and residents are proud of its appearance and the way it fits its surroundings. Even many of those who originally opposed the project are happy with the result.

True to the ideals of structural art, the bridge achieves its aesthetic impact through the careful shaping of the structural members themselves to both efficiently carry the forces on them and create a memorable visual impression. It is a masterpiece of the bridge builder's art.

The bridge was designed by the staff of the California Department of Transportation. Kelly Holden led the design team.

RIVERS AND TIDAL WATERWAYS

This section addresses bridges that cross large water bodies. Small waterway crossings have much in common with highway overcrossings, or they may be incidental to a major structure over a valley, both of which have already been discussed. While there obviously is a continuum of waterway sizes, this section focuses on sites where the water surface is a dominant feature in the visual scene (Figure 6-31).

Developing a Design Intention/Vision

Physical Requirements

The width, depth, navigational requirements, and bottom conditions of the waterway will be the overriding site conditions for this type of structure. The presence of a navigational channel will probably be the major consideration when setting the height and length of the largest span. The nature of the foundations required and their cost will largely determine the economical span range for the structure. Submerged utility lines may be a concern, and there may be a need to carry utility lines on the structure. Frequently there are railroads or roadways on the shore that must also be spanned. Roadways often intersect with the bridge at or near the shoreline, creating a need to gracefully widen the structure to accommodate intersecting ramps and streets. These bridges are often called on to provide pedestrian and bicycle access, and access for fishing, all of which need to be anticipated from the beginning.

Waterway bridges typically have one or a few main spans. Bridges across wide waterways will have a series of repetitive approach spans that can comprise a much larger portion of the total bridge than the main spans.

Visual Environment

The visual environment is first established by the width of the waterway. Very wide waterways (wider than one mile) will themselves be the dominant aspect of the visual environment. For narrower waterways, the nature of the shoreline will be

FIGURE 6-31 *With very simple elements, this Tennessee structure conveys strength, continuity, and horizontality. The last river span is tapered to make a smooth transition between the depth of the main spans and the depth of the approach spans. (See color insert.)*

more important in setting the visual environment. The range of shoreline uses can be very broad, including industrial, commercial, residential, park, or natural. At any given site one or two uses usually dominate. An example would be industrial uses on one side and a park on the other. The visual effect of the uses should be confirmed in the field with photographs. An industrial use may in fact be screened from view by riverbank vegetation, such that the actual visual environment is quite natural. The resulting information should be used to make key decisions related to the visual environment.

Nearby and Associated Uses

Public shoreline open space, such as a riverfront park, fishing pier, or festival marketplace, will be very important in determining viewpoints and the nature of the design. Commercial or residential uses offering a number of potential overlooks will also be important considerations. Industrial uses offering no public access will require less consideration.

Symbolic Functions

If the bridge is at or visible from a town or neighborhood center, it will become, for better or worse, both a landmark and the entrance gateway to the town or neighborhood. If done well, it may become a symbol of the town.

Boundaries

Sometimes these bridges end conveniently at the shoreline. More often the boundaries are not so clear. The project may extend across adjoining rail lines, roadways, and riverbank areas. A riverfront walk may adjoin the structure and require consideration as to detail and connection. There may be a desire, for example, to carry the same ornamental lighting that is used on the river walk onto the bridge.

Anything that can be seen at one time from the important viewpoints should be considered part of the structure for design purposes. The goal should be to develop a consistent concept that carries across the entire length of the project.

Guidelines for the Design Intention/Vision

The most important views of bridges over rivers and tidal waterways are usually oblique views from points along the shore. Particular locations can be identified that are likely to be the most-favored viewing points. For example, there are often areas along the shore dedicated to public docks or parks, where people are likely to congregate. A shoreline condominium development will have many observers on balconies. In the event there is a curve in the roadway approaching the bridge, drivers may also get a view of the bridge prior to crossing it. On waterways with a significant amount of recreational boat traffic, the view from the water should also be considered. The point of contact with the shore can also be an important feature in forming the visual impression, particularly if there is pedestrian use or a shoreside roadway at that point. Photographs should be taken from the most important viewpoints and used as the base for future sketches and renderings.

At the most likely oblique viewpoints, the most important features in determining the visual impression will be the shape of the basic horizontal and vertical geometry, followed by the structural type or types, and the transitions between different

FIGURE 6-32 *Both horizontal and vertical curves are present, but they blend into an uninter-rupted flowing line.*

FIGURE 6-32 *Both horizontal and vertical curves are present, but they blend into an uninter-rupted flowing line.*

types or shapes. Pier shape is the next most important consideration, particularly as seen at an oblique angle with piers lining up one behind the other. Superstructure depth may be of less importance, but the apparent thinness resulting from the depth-to-span ratio is still important. Finally, parapet and railing design will be important to viewers on the shore close to the bridge and will be a very important feature to users of any sidewalk or bike path. Exact pier and abutment place-ment, as well as abutment shape, are generally of less concern for these structures.

Bridges of this type can be quite long and are usually seen in their entirety from the most important viewpoints. It is the whole of the bridge that will make the visual impression. When designing a bridge across a river or tidal waterway:

- Aim for a display of the basic geometry as a single sweep of roadway "ribbon"—one unified form.

- Keep the spans as long as possible to keep views open and to minimize the "forest of piers" look.

- Keep pier shafts as thin as possible to keep the structure transparent.

- Emphasize the horizontality of railings and parapets; avoid pedestals, pylons, and pilasters that break the horizontal sweep of the structure.

The goals at the shoreline should be focused on the uses at that point. Pedestrian or park uses under the bridge, for example, may suggest a concern for the appearance of the underside. The need for access to a pedestrian walkway may impose a need for gradual slopes for a ramp or stairway.

Determinants of Appearance

Geometry

The topography near a river crossing is usually obvious. People can understand it easily by the shape of the waterline and the profile of the adjoining bluffs. People will judge the appearance of these bridges by how logically they fit the topography—in particular, whether they are aligned along the shortest apparent distance between shores.

Because these bridges can be quite long, and the whole bridge will be visible from many viewpoints, the horizontal and vertical geometry will be critical in forming its visual impression. Any flaw will appear as an obvious kink. This visual effect should be considered from the beginning in establishing the horizontal and vertical alignments. The goal is to construct both vertical and horizontal alignments from long continuous curves. The overall appearance should be one continuously flowing "ribbon" in space (Figure 6-32).

Superstructure Type

This type of structure often requires several different span lengths to address varying requirements at different points along the structure, resulting in different superstructure types, sizes, or shapes at different points in the structure. It is not always easy to make a clear and simple transition between them, but doing so is critical to the visual success of the structure. The bridge should be a single unified concept; changes in structural type or depth to accommodate differing span conditions should be made smoothly; similar structural shapes, materials, and/or colors should be used to tie the structure together (Figure 6-33).

Pier Placement

In the oblique views in which these bridges are usually seen, the piers will appear to line up behind each other. Thus, the number of piers becomes a significant visual parameter. The fewer piers, the less wall-like the bridge will appear when seen in the oblique view. Determining the number of piers and pier spacing is typically seen as primarily an economic issue. Decisions are often made by considering superstructure and pier costs separately, considering each based on rules of thumb and experience from other bridges. This will produce misleading results. Superstructure costs and foundation costs should be considered together when developing optimum span lengths, as described in Chapter Four. Spans at the long end of the minimum range should be chosen. The bridge in Figure 6-31 (page 224) has a graceful combination of river spans, but the land spans seem too short.

For arch and rigid-frame bridges, the placement of the springing point should be on-shore near the shoreline to give the structure a strong visual end point. See Chapter Four for additional guidelines.

Abutment Placement

The shoreline is the obvious place for the abutments, which can be fine as long as they are not too high. More likely the presence of floodplain or shoreline land uses will require that the abutment be placed some distance from the shore. Then the placement will be a matter of accommodating those uses (Figure 6-34). If there are no shoreside uses, the final consideration will be the desirable height and lengths of approach fills. In that case the structure will look best if the abutment has some clear relationship to topographic features. For example, locating the abutments at riverside bluffs usually makes sense economically and will be visually logical.

FIGURE 6-33 *This Charleston, South Carolina, structure smoothly accommodates the long main span simply, by deepening the same trapezoidal box used for the approach spans; the structure is so long that the piers still line up into a wall at this oblique angle, even with long spans and a simple pier design. A paired-column pier design similar to the one shown in Figure 6-27.1 (page 219) might have kept the structure more transparent.*

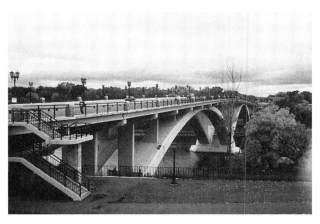

FIGURE 6-34 *The abutment of this Minneapolis, Minnesota, bridge allows for shoreline uses.*

Superstructure Shape

Accommodating the various spans required is the main design challenge of this type of structure. If all the spans can be accommodated with girders, the matter can be simplified. The structural depth can be kept constant or smoothly varied over the entire bridge. The tapered river bank spans of the bridge in Figure 6–31 (page 224) gracefully accommodate the change in depth from main span to approach spans. Haunched girders offer a visually logical way to accommodate longer spans while still maintaining uniformity of depth with adjoining spans. The haunches visually punctuate the sweep of the bridge and draw attention to the longest span.

Deck overhangs are important for both the superstructure and the substructure. The deck overhang should be continuous throughout the structure as a way of emphasizing the visual continuity of the bridge (Figure 6-35). The shadow accentu-

FIGURE 6-35 *The wide overhangs of Oregon's Alsea Bay Bridge create a shadow line that ties this structure together.*

FIGURE 6-36 *These visually complex piers seem to hold down and break the smooth sweep of the girder.*

FIGURE 6-37 *By comparison, the piers of the Pescadero Creek Bridge in California appear as extensions of the curves of the girder. This is another view of the bridge shown in Figure 2-57.*

ates the continuity of the superstructure, and the overhang permits narrower piers with fewer columns, producing a more transparent bridge.

Parapet faces and railings will be important in establishing the visual impression of this type of structure. The parapet and any railings should be used to reinforce the horizontal continuity of the superstructure. Since small details will not register on bridges of this size, the design should use large offsets, insets, or repetitive patterns that emphasize horizontal continuity. The simple railing design on the bridge in Figure 5-45 (page 172) allows the horizontality of the structure to dominate.

Pier Shape

In the oblique views in which these bridges are usually seen, the piers will appear to line up behind each other. The more elements in each pier (columns, pier cap, etc.), and the more angles at which they are placed, the more confused the whole effect will be. If possible, use no more than two columns/piers at each pier line. If more columns are necessary for unusually wide structures, they should be paired. (See Figure 6-27.1, page 219.)

The structure will be seen from many viewpoints, so it is important to design piers that create pleasing compositions when seen together. Keep the slopes and curves of adjoining piers consistent. (See Figure 5-22, page 154.) If the structure gets higher or wider, it will be necessary to provide a method of gracefully lengthening or widening the pier so that the series of piers looks good when seen together.

When haunched girders are used at main channel spans, the main span piers will have a different shape from the flanking piers. Arch, truss, and rigid-frame structures may also require special pier shapes. Because most or all of the piers are visible at once in the oblique view, it is important that these piers also display a continuity of form and shape.

Just as for the bridges discussed in the previous section, the pier cap will be a prominent element in the oblique view, and will interrupt the visual lines of the structure unless it is minimized. The designers of the bridge in Figure 6-31 (page 224) deals with the problem by making the horizontal dimension of the pier cap end

FIGURE 6-38 *These piers, from the bridge in Figure 6-31 (page 224), develop the sense of strength appropriate to the long river spans they support; the same piers under the much shorter land spans, however, seem heavy.*

seem larger than the vertical dimension (Figure 6-38). The end is split horizontally into two planes, one angled skyward and the other angled toward the water. See Chapter Five for other ideas on minimizing the pier cap end.

A clear demarcation between superstructure and substructure at the bearings will help the structure seem to stand free and emphasize its horizontal continuity. Bearings and pier tops can be detailed to provide a visual demarcation, as outlined in Chapters Four and Five (see Figure 5-25, page 156).

Abutment Shape

With this type of structure, the abutments should be shaped to respond to their immediate surroundings. If there are pedestrian uses nearby, the abutment may be an important point of access to walkways on the bridge. Or, it may be necessary to relate the design of an abutment to the design of a nearby river walk or the architecture of a nearby building (Figure 6-39). If there are no nearby associated uses, the abutment can be kept simple. Because of its relative size compared to the structure as a whole, it will have little effect on the appearance of the bridge.

Color

Color should be used to reinforce the horizontal continuity of the structure. Similar elements, such as approach girders and arch stringers, should be painted the same color across the whole structure (Figure 6-31 page 224). Where steel and concrete girders are used together, the steel paint color should be selected to match the concrete. Bright color applied to a secondary element, such as a railing, can be used to visually pull together different structural types. The bridge in Figure 6-39, with its light-colored parapet, and the one in Figure 5-45 (page 172), with its distinctive railing color, show the effectiveness of using a colored horizontal element to unify a structure and emphasize its horizontal sweep.

When selecting colors, the apparent color of the water will be an important consideration over the wider waterways. Apparent water color is a function of the color of the water itself, which can vary by season depending on runoff and temperature, and sky color, which will also vary by day and season. Multiple observations from the important viewpoints over an extended period will be necessary to establish the dominant color or colors that should be used as a basis for the decision.

Surface Patterns, Texture, Ornamentation, and Details

Traditional brick and stone are generally not convincing when used as superstructure components of these bridges. The spans of these bridges are generally so much larger than historic masonry structures that the use of the materials is not credible. Stone or brick pier facing is one use of traditional materials that is consistent with historic uses of these materials.

FIGURE 6-39 *The abutment and sidespans on the Discovery Bridge in Columbus, Ohio, incorporate provisions for a riverwalk.*

Most of the bridge will be in view over the water at one time and from many viewpoints. Drawing attention to points of interest along the structure can improve the appearance of these structures. Structural elements can be designed to perform this function by adding color or detail at points of structural importance, such as bearing stiffeners in a fan pattern at the piers of a steel girder bridge. If there is a pedestrian walkway on the structure, overlooks at key piers will punctuate the sweep of the structure and provide a pedestrian amenity.

Pedestrian walkways on river crossings must be made wide enough to be comfortable for two people to walk abreast or pass one another. Because of the length of these structures, an otherwise adequate width can seem narrow. If the decision is made that there will be enough users to make a walkway on the bridge worthwhile, then it ought to be wide enough to be comfortable for those users: 6 feet is too small; 8 feet should be the minimum, with 10 feet the minimum for walkways that are expected to get significant use or bicycle traffic. Patterns in the walkway surface created by scoring or color in the concrete or by changes in materials should be used to visually break down the apparent length of the journey into manageable increments.

Signing, Lighting, and Landscaping

Because of their length and the frequent presence of intersecting roadways at their ends, this type of structure is often called on to support sign structures. Integrating the sign structure with the superstructure and/or parapet design is the important factor. It is important to place them in some logical relation to the overall bridge, such as at a pier, and to smoothly integrate the connection of the support structure into the overall structure. The sign structure should seem like part of the original concept, not a bolted-on afterthought.

Lighting is also often a consideration for this type of structure because of the length. Size, spacing, and style of the fixtures may have a significant impact on appearance, and should be considered in relation to the overall design intent for the bridge. The fixtures will produce reflections in the water that will magnify their impact at night. Sharp cutoff fixtures can reduce the overspill of light. Pole spacing should coincide with, or be an even multiple of, span or the spacing of other struc-

FIGURE 6-40 *The lights on this bridge are spaced to fit into the spans (six per span) and are coordinated with the railing post spacing as well.*

FIGURE 6-41 *Ingraham Street Bridge over Mission Bay, San Diego, California. Bridge light reflections on the water are an important part of the nighttime scene.*

tural features. If the lighting is to continue along an adjoining roadway or river walk, that light fixture and pole spacing should be consistent with the overall concept and detailing of the bridge lighting (Figure 6-40).

Bridges over water in prominent locations can be very effectively lit at night to create a night-time landmark. Reflections from the water multiply the effect. The most effective approach is to light the piers and/or underside of the structure, which takes best advantage of the reflections (Figure 6-41). Colored light can add an additional dimension. The options are many, but there are also many areas where things can go wrong. This is definitely an area for which to engage a lighting specialist.

Landscaping elements are too small to have much of an impact on the appearance of the overall bridge. However, landscaping may be a very important part of shoreline uses, and may affect the appearance of bridge elements on-shore, particularly abutments.

CASE STUDY

Clearwater Memorial Causeway, Clearwater, Florida

The Clearwater Memorial Causeway bridge crosses Clearwater Bay, which separates downtown Clearwater, Florida, from the barrier island of Clearwater Beach. The beach is both a retirement and recreational destination, attracting large volumes of traffic in peak seasons. Operation of the existing four-lane drawbridge backs up traffic into the downtown area and onto Clearwater Beach. The city decided to replace the drawbridge with a high-level bridge to eliminate these backups.

The physical requirements for the new bridge were to carry four lanes of traffic, a left-turn lane at the city end, and two 12-foot sidewalks, one of which is the route of the Pinellas Trail, part of a countywide bikeway system. The bridge provides 74 feet of vertical clearance, which exceeds the height of almost all recreational boats using the area, over a 125-foot-wide navigational channel. To reach the additional height, the city approaches curve and tie into a street two blocks south of the existing bridge terminus. On the beach side, the approaches extend about 1,500 feet along the causeway beyond the existing bridge terminus. The visual environment is striking. It is dominated by the sweep of Clearwater Bay, which is bordered by single-family residences and mid-rise condominium buildings. Views toward the beach focus on the existing Clearwater Pass Bridge and the Gulf of Mexico beyond. The sunsets are often spectacular, and the city markets the view itself as a tourist attraction. The existing bridge was supported on 14 closely spaced pile bents, with a total of 56 piles. When observed from the shore, the pile bents and the bascule piers effectively combined into a visual wall, blocking views up and down the bay (Figure 6-42).

FIGURE 6-42 *The previous Clearwater Memorial Causeway Bridge, Clearwater, Florida, as seen from Coachman Park. The distant buildings of Clearwater Beach could barely be seen between its piers and piles.*

On the downtown side, the bridge lands at the city's waterfront park, Coachman Park, and its civic center. City hall, the city's main library, and the convention center adjoin the park and overlook the bridge site. The park is the location of the city's band shell, and is the site of frequent outdoor festivals and concerts, which often extend into the evening hours.

The ground surface of the waterfront area is now mostly occupied by the streets approaching the existing bridge. With the completion of the new high-level bridge with its access point several blocks away this space will be available for other uses. The city is planning various extensions of the park and civic uses presently bordering the site. The city is also attempting to enhance the attractiveness of the beach to visitors and residents, and to similarly enhance the attractiveness of the downtown itself as a destination for recreation and shopping. Given its location in the very center of the city's civic space, the city wanted the bridge to become a positive symbol of the city. Because the bridge is at a much higher elevation than the existing bridge, the boundaries of the project extend for a much greater distance, to the point at which the profile returns to existing grade. On the city side, this extends several blocks from the point at which the bridge crosses the shoreline.

In order to determine the public's ideas about the project, the city organized a community committee that had direct input into the study at key milestones. The design intention/vision that resulted from this process focused on preserving and enhancing the view from the waterfront, both during the daytime and evening hours. A second consideration was to enhance the passage across the bridge for beach-bound travelers. A third requirement was to support and enhance future use of the area along the waterfront under the bridge. Finally, the city was looking for a design that would be a suitable landmark for the city, but that could be built for the money available from state and federal agencies and a local tax initiative.

The geometry of the bridge was set by the need to cross the channel at a high elevation and connect to the existing street system in Clearwater and the existing causeway on the Beach side. With those requirements in mind, the horizontal and vertical curves were set as long as could be accommodated.

The structural type and pier placement were considered together as part of a study of the visual impact of the bridge. A view corridor was identified, covering the sight lines from the most important view points in downtown Clearwater (Coachman Park, the convention center, and city hall) to Clearwater Beach and the Gulf of Mexico. The views from these locations intersect the center half of the bridge (Figure 6-43). Simply spanning the navigational channel would require spans on the order of 150 feet, which could be accomplished with precast concrete bulb tees on hammerhead piers. However, this would place 11 spans in the primary view corridor, with 12 two-column piers—a total of 24 vertical shafts in the view. This option was compared with one using 300-foot spans that cut in half the number of pier shafts in the view corridor. The effects on the view were illustrated for the community through a series of three-dimensional renderings. The city commission members considered the options and rejected the 150-foot spans because of their effect on the view. They then allocated the $2.5 million required for the longer spans in the view corridor. People often ask what improved aesthetics are worth. In this case, preserving their view was worth $2.5 million to the people of Clearwater.

The design team investigated a series of long-span arrangements, with spans up to 500 feet, for their costs, constructability, and their effects on the view corridor. The options were illustrated with computer-drawn renderings placed on photographs taken from the convention center and the park shoreline. The city concluded that a plan using three 360-foot spans for the center part of the bridge, flanked by two 280-foot spans, produced the most cost-effective combination. This reduced the pier lines in the view corridor to three, with a total of six vertical shafts (Figure 6-44). The remaining piers are outside the view corridor and have spans varying from 120 feet to 196 feet. The abutment on the city side was placed to provide space for the access road to a waterfront condominium building and on the causeway side to achieve a reasonable end height (30 feet) for the abutment retaining wall.

With this span range, and considering the marine location, a trapezoidal concrete box structure was selected as the superstructure type. Haunches are located at the four main span pier lines, giving the bottom edges of the girders a continuous curve in elevation. The variation in box depth, combined with the trapezoidal cross section, cause the soffit of the girders to vary in width. These curvatures, when combined with the curved horizontal and vertical alignments of the bridge, give the structure a flowing, undulating appearance. Overlooks are provided at the main span piers to give pedestrian users a place to rest and enjoy the view..

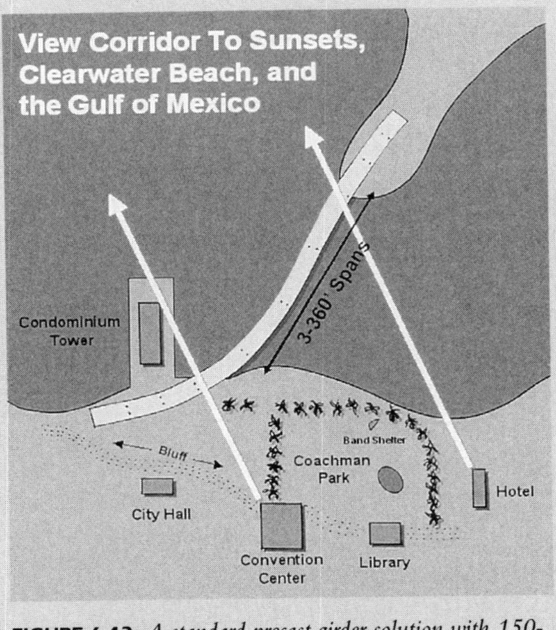

The piers are shaped to minimize the shaft thickness at water level. They are an elongated octagon in plan, giving them a faceted appearance, which makes them seem thinner and taller. At their tops the piers flare outward to provide space for the bearing and to give the pier edges a curvature that is commensurate with the curvature of the girder haunches. The abutments are created by mechanically stabilized earth walls faced with concrete panels. The walls continue for some distance along the approach roadways to minimize the footprints of the approach embankments. The bridge railing and parapet are carried along the tops of the walls to provide a coping and to tie them into the bridge. The concrete is coated with an off-white color to keep the bridge as reflective as possible. Depending on the time of day and the seasons, the strong Florida sun causes the bridge to take on a hint of the colors around it, particularly the light green of the bay. The railing and light posts are painted a medium blue, tying them into the sky against which they are usually seen. The concrete panels have a repetitive vertical groove

FIGURE 6-43 *A standard precast girder solution with 150-foot spans would have required 12 two-column bents, for a total of 24 vertical columns in the view corridor. By using 360-foot spans, the number of columns in the view corridor was reduced to six.*

intended to obscure the joints of the repetitive wall panels.

The bridge has no large signs. The roadway lighting is placed in the median, which cuts in half the required number of poles and creates an obvious visual order. The poles are high enough so that the roadway fixtures also light the walkway surfaces. Blue floodlights are mounted on the pile caps, aimed to light the soffits of the main spans, creating an abstract wavelike

FIGURE 6-44 *Clearwater Memorial Causeway, Clearwater, Florida. Long spans open up the view to Clearwater Beach and the sunset (see color insert).*

form that frames the nighttime view of Clearwater Beach (Figure 6-45). The lighted soffits also form a lighted nighttime ceiling for the waterfront spaces beneath the bridge, making that area pleasant at night as well.

With the long spans, slim pier shafts, and planar, light-colored box girder surfaces, the bridge reopens the view corridor that was shut off by the existing bridge, providing clear views to Clearwater Beach and the Gulf of Mexico. The space below the structure is open and light-filled during both daylight and nighttime hours. Waterfront visitors find it a pleasant place to be. Figure 3-26 (page 89) shows the bridge under construction and provides an indication of how open the view will be following the removal of the existing bridge.

The design of the Clearwater Memorial Causeway Bridge was led by HDR Engineers, Inc., with Theun van der Veen as the design engineer. The author provided aesthetic advice, urban design, and community participation services, through his firm, Rosales Gottemoeller & Associates.

FIGURE 6-45 *Blue floodlights will make the girder soffits a frame for the nighttime view, an effect multiplied by shimmering reflections in the water (see color insert).*

INTERCHANGE RAMPS AND VIADUCTS

Simple diamond and cloverleaf interchanges usually have one structure or paired structures that are basically highway-over-highway bridges, hence can be treated as described in the "Highway-over-Highway Bridge" section, earlier in this chapter. More complex interchanges require multi-span ramps. They form a separate category, with many features in common with viaducts.

Developing a Design Intention/Vision

Physical Requirements

Ramps and viaducts typically have a complex set of requirements related to the undercrossing roadways, railroads, utilities, and land uses. Each of these will have horizontal and vertical clearance requirements, some of which may overlap or conflict. The first step is to locate all of the requirements on a plan and profile of the structure and see whether any logical pattern presents itself. The challenge of the structure will be to develop an arrangement of piers and superstructure that creates an apparent visual order among the conflicting requirements.

Visual Environment

The visual environment for ramps is usually dominated by the highway interchange of which they are a part. The graded, landscaped slopes and other nearby bridges will set the backdrop against which the structure must make its impression. These environments are spread out, but visually complex, with curved roadways seeming to intersect in space in ways that may not be immediately obvious. Drivers have the problem of finding their way through the complexity, a stressful process that distracts from their ability to enjoy the experience.

Viaduct sites usually have a different kind of visual complexity. The undercrossing roadways, utilities, and other elements are often at varying angles to each other and the structure. Even if the undercrossing facilities are on a street grid at right angles to the structure, there will be buildings, stands of trees, parking lots, and vacant spaces of various sizes and shapes and at varying intervals. Again, the challenge is to bring some apparent visual order from the complexity.

Nearby and Associated Uses

For a ramp, the nearby uses are mostly highway-related. The bridge must provide for safety for off-the-road vehicles and sight distance to ramp terminals and the traffic ahead. Ramps sometimes are part of shopping malls, office parks, or even downtown development, in which case buildings become part of the backdrop of the structure. Their uses, size, color, and materials should become a consideration in determining the appearance of the structure.

Viaducts are usually in and among other uses. The appearance of nearby buildings and natural areas, and the uses expected for the area under the structure, need to be considered in forming the design intention/vision. Areas under viaducts are often used for pedestrian circulation, parking, parks, and even industrial and commercial uses. The underside of the viaduct then becomes the "ceiling" of a large outdoor "room," thus its appearance and lighting become important considerations (Figure 6-46).

Symbolic Functions

Ramps are usually seen as a functional feature of the larger interchange of which they are a part. It is difficult and unusual for them to have a symbolic function apart from that of the interchange itself. Major ramps can contribute to the attractiveness of interchanges by symbolizing the smooth and efficient flow of traffic. Major interchanges are sometimes seen as milestones along a freeway, as points of orientation or

FIGURE 6-46 *This viaduct visually organizes a complex scene and creates the "ceiling" of an outdoor "room."*

entrance. As the most prominent feature of the interchange, the ramp structures might be called on to demarcate the interchange or set some theme. That can be accomplished by picking an unusual structural type, parapet or abutment pattern, or color, and using that feature on all of the structures of a given interchange.

The symbolism associated with viaducts is often negative. It was a viaduct that first earned the nickname "the Chinese Wall." Viaducts can be seen as the dividers of communities and the attractors of building vacancies and vandalism. Avoiding these effects depends on maintaining clear sight lines through the structure in as many directions and from as many viewpoints as possible. It also depends on arranging piers in logical patterns that complement the pattern of the underlying facilities. It is important to avoid the creation of odd bits of land and hidden areas that cannot be reused and that become unattractive nuisances.

Boundaries

Ramps should seem to blend into the geometry of the roadway they are carrying, to become an extension and continuation of that roadway. In that sense, they should not have a clear boundary.

Viaducts are generally long and near other uses, often making it difficult to see the entire structure at one time. All that can be seen is a slice of the structure. Again, the structure does not appear to have boundaries. However, the boundary assumed along a viaduct can have a significant effect on the visual quality of the result. Particularly important is the condition of the ground surface under the viaduct. If the boundary of the structure is assumed to be the bottom of the piers, then the ground surface will be left bare and unkempt, and the structure will attract the nuisances just described. The boundaries should be set at the edges of the ground plane of the nearest uses, and the whole surface area between those boundaries developed as part of the design of the structure. That may mean paving for parking, landscape or hardscape for park uses, or whatever is required to bring the underbridge area back into a productive use that complements its neighbors.

Guidelines for Developing a Design Intention/Vision

The viewpoints in complex interchanges are usually a multiplicity of locations along the intersecting roadways. The most important can be identified based on traffic volumes of the roadways from which the ramp will be visible, or length of time a particular structure is in sight from a given roadway. If the ramp is among or near adjoining buildings, then points of visibility where people gather in or among the buildings should be located.

Viaducts, particularly in urban areas, will have an almost infinite number of view points, many of which may involve pedestrian traffic in close proximity to the bridge. With this range of possibilities, it is hard to identify specific features of the bridge as being more important than others. Any and all features could be important depending on the circumstances. However, with both ramps and viaducts, certain features will be important in all cases.

Because these structures are long and often seen a oblique angles, the shapes created by the basic horizontal and vertical geometry will strongly influence the visual impression. Any flaw will be apparent as a jarring kink. Pier placement relative to the underbridge area will be critical. It will affect safety around interchange ramps and the viability of underbridge and nearby uses at viaducts. Selection of structural type is important because it will determine the span lengths available, the appearance of the underside, and pier configuration. Superstructure depth-to-span ratio, to create a sense of thinness, is important. Finally, parapet and railing design will be important to viewers close to the bridge and will be a very important feature to users of any sidewalk or bike path on a viaduct. Exact abutment placement, and abutment shape are generally of less concern for these structures.

The most powerful visual aspect of interchange ramps and viaducts is the sweep of the roadway geometry itself. It is the most important visual characteristic of the structure and must be enhanced to develop the visual strength necessary to overcome the surrounding visual cacophony. The lines of the superstructure should parallel the basic geometry as much as possible so as to reinforce it. The goal is to create a structural ribbon carrying the roadway over and under various obstructions. The second most powerful element is the placement of the piers in relation to the sizes, spacing, and alignment of the underlying facilities. The third most important characteristic is how the bridge integrates with other nearby bridges and retaining walls, which are often present in interchanges and urbanized areas. Because of the need to focus attention on traffic signals and signs, and because of the frequent presence of pedestrians, the design of the underside is important.

Urban interchanges and viaducts are inherently confusing, therefore the visual design goal should be to make the structure and appurtenances as simple as possible. Bridges of this type can be quite long, and are rarely seen in their entirety from one position; one's impression is formed by a succession of views from the most important viewpoints. It is important that all of these views combine into a consistent and easily understood whole. Once again, simple details and consistent shapes used throughout the structure and aligned along the outlines of the major structural elements help to create the impression of visual order. When designing a ramp or viaduct:

- Aim for a display of the basic geometry as a single sweep of roadway "ribbon"—one unified form.

- Make the spans as long as possible to keep views open and minimize the "forest of piers" look;

- Establish a strong and logical order that complements the patterns of adjoining and under-bridge uses.

- Keep pier shafts as thin as possible to keep the structure transparent

- Focus on a few consistently used materials, a few consistently used shapes, and keep surfaces and shapes continuous.

- Emphasize the horizontality of railings and parapets; avoid pedestals, pylons, and pilasters that break the horizontal sweep of the structure.

Determinants of Appearance

Geometry

The length and position of these structures means, their geometry will be the most powerful element in forming people's visual impression. That shape will be visible from many of the important viewpoints, therefore it is critical that the geometry produce an attractive overall shape, without dips and kinks. The horizontal and vertical alignments should be constructed from long, continuous curves. The geometry will also determine the nature of the structure alignment with undercrossing roadways and uses. Adjustments in the alignment that would make possible a more regular placement of piers or avoid the creation of leftover slivers of land should be investigated.

Superstructure Type

Girders are the most frequent form for ramps and viaducts, as the spans are generally not long enough to require other types. Girders provide a ribbon of structure that fits the need for visual continuity. Relative slenderness should be sought when picking superstructure type, and curved girders should be used for curved roadways.

Because of the height and locations of these bridges, their undersides will often be prominent, particularly where pedestrians are involved. Pedestrians will read the underside of the bridge as the ceiling of an outdoor room, thus the designer should design the features of the underside to be seen, because they will be, and they will be important factors in forming people's visual impression of the bridge. Concrete slab and box structures avoid visual clutter and provide a light-reflecting surface for the ceiling. Figure 6-46 (page 237) is a good example. Steel box structures can have similar qualities, particularly if they are painted a light color. Box girders with their interior pier caps also provide flexibility in pier placement, which can help resolve conflicting underbridge conditions.

Precast I-girders and steel-plate girders present more of a challenge because they usually require piers with a visible cap and because of their unattractive undersides. Viaducts and ramps often have to contend with many conflicting geometries in undercrossing roadways, railroad tracks, and utilities, all passing beneath the structure at different angles (Figure 6-47). The common solution is to arrange the pier caps at differing angles to the roadway geometry in an effort to mediate between the roadway and what it must cross. The visual result of this approach is always disastrous (Figure 6-48). What is already a visually complex situation becomes hopelessly confused as the pier caps interrupt the roadway geometry in differing ways and disrupt any chance for the roadway to create an overriding order. The structural situation becomes more complex, too, as every girder and pier cap becomes a different length.

FIGURE 6-47 *A good ramp structure like this one in Reno, Nevada, becomes a visual reflection of the roadway geometry itself.*

FIGURE 6-48 *Three different pier types cause visual confusion.*

Integral pier caps are a potential solution. They are less visually prominent, so their alignment does not interfere visually with the roadway geometry. They also provide flexibility in pier placement, which can help resolve conflicting underbridge conditions while allowing a uniform girder span.

The undersides of precast I-girders and steel-plate girder bridges are unattractive because of the dark, shadowed areas between the girders and the visually confusing and dirt-catching bracing required. Such bridges can be made acceptable by simplifying the details, providing under-bridge lighting, and, in the case of steel, using a light paint color. See Chapter Five for a discussion of details.

Pier Placement and Type

The placement of piers offers a major opportunity to create a visual order that will organize a complex scene. A strong and logical pattern is the goal. The best approach is to plot all of the restrictions on pier placement on the plan view—roadways, pedestrian areas, utility crossings, buildings, and streams—then develop pier placements that satisfy the restrictions and still preserve a clearly organized layout. Piers should be placed in a consistent relationship to undercrossing streets and other features so as to create a regular spacing or a smoothly varying progression of different spacing. Undercrossing roadways at differing angles will suggest piers skewed at different angles. This will result in a complete confusion of appearance from below. Figure 2-8.1 (page 31) is an example of such a result.

To avoid this problem, piers should be placed so they are parallel to others in a series or radial to the bridge roadway. The worst problems are caused by forcing a superstructure with numerous parallel I-girders onto a site with complex conditions below. This can't help but result in a variety of piers cobbled together to meet all of the different clearance requirements (Figure 6-48). That situation is a clear indication that the wrong type of superstructure has been chosen. A box girder structure that allows fewer, narrower piers and flexibility in their placement should be used instead (Figure 6-49).

The structure will often be seen from oblique angles, so the piers will often seem to line up one behind the other. There are two ways to alleviate this problem. The first is to extend span lengths, considering superstructure costs and foundation costs together to develop optimum span lengths. The second is to use wider overhangs in the superstructure to allow narrower piers. Both approaches open up spatial corridors

FIGURE 6-49 *A torsionally stiff box girder for the Whitehurst Freeway in Washington, D.C., allows thin, widely spaced central piers, freeing space and sight distance for complex roadways below.*

FIGURE 6-50 *Long span lengths and wide overhangs with narrow piers allow diagonal views through Christian Menn's viaduct in Felsenau, Switzerland.*

through the structure, including the diagonal corridors often seen from important viewpoints (Figure 6-50).

Abutment Placement

For ramp structures, the abutment placement should begin with the grading plan of the interchange as a whole. The slopes should be shaped to provide smoothly flowing landforms. Once that is accomplished, the abutments can be established. Abutment heights should be kept to a minimum, to emphasize the horizontal continuity of the structure. For viaducts, abutment placement will be a function of the location of the underlying facilities that are being spanned. Abutment placement should be approached using the same guidelines as those used for pier placement. The abutment should be seen as simply the last pier in the series.

Superstructure Shape

Superstructure shape should be developed to emphasize the continuity and visual sweep of the structure. It is important to seek girder dimensions and details that emphasize apparent thinness and horizontality, and to keep structural depth constant or smoothly varied over the entire bridge. Haunches can be used on a ramp or viaduct for a few unusually long spans. However, a long series of haunches tend to look disturbing unless the spans are more or less equal and the height above the

FIGURE 6-51 *The continuation of the parapet overhang onto the abutment extends the horizontal shadow line of the structure.*

FIGURE 6-52 *Viaducts and ramps should split in a simple and clear manner.*

ground plane is similar from span to span. Keeping the parapet-girder-overhang edge condition uniform along the structure will reinforce the sweep of the roadway geometry, on which the visual success of the structure depends (Figure 6-51).

Ramps and viaducts often require widening and splits in the bridge to accommodate ramps and acceleration lanes (Figure 6-52). Girders should be laid out in such a way as to logically accommodate the gradual change in width. The parapet face can be shaped to emphasize the horizontal continuity of the structure by means of offsets to create shadow lines and other patterns. Railings and noise walls, if present, should be detailed with the same goal in mind. The horizontal elements should be visually the strongest; and pedestals, pylons, and pilasters should not extend above the top edge of the dominant horizontal element, whether railing, fence, or noise wall.

Pier Shape

Because there will be so many piers, and because they will be visible from so many different view-points, it is important to minimize the number of elements and keep the shape simple (Figure 6-53). One way to do that is to use piers that elimi-

FIGURE 6-53 *Large overhangs emphasize horizontal continuity and minimize pier width. The notched piers create "fingers," which seem to barely touch the superstructure floating overhead. (Note how the girder smoothly deepens to provide a haunch for the longer span in the distance.)*

nate or minimize the pier cap. This is particularly important when undercrossing streets cross at various angles, as the pier caps will be at conflicting angles with all but one of the undercrossing streets, thereby setting up a visual conflict.

Keeping in mind the structure will be seen from many viewpoints, it is important to design piers that create pleasing compositions when seen together. Keep the slopes and curves of adjoining piers consistent. If the structure gets higher or wider, it will be necessary to provide a method of gracefully lengthening or widening the pier so that the series of piers looks good when seen together (See Chapter 5 for examples, as well as Figure 6-52.)

Abutment Shape

Ramp abutments in landscaped interchanges will be relatively minor elements in the visual scene and should be kept simple, small, and shaped to emphasize the continuity of the ramp structure. If a ramp abutment ties into a retaining wall, it should be detailed to carry through the same pattern as the adjacent retaining wall. Viaduct abutments will often be prominent elements from nearby viewpoints and should be compatible with surrounding features. If there are buildings or particular street features nearby, the abutment can be detailed to reflect them.

Color

Color can be used to emphasize the continuity of these structures. Placing a strong color on the girder, railing, or some repetitive feature of the parapet will tie the whole structure together and give it a presence that will help the structure unify the overall scene.

Surface Patterns, Texture, Ornamentation, and Details

A continuous surface texture or pattern on the parapet face can also help unify these long structures. The abutments are such small parts of most viaducts that textures or patterns on them will not have much effect on the overall impression. That said, abutment patterns or textures can improve the impression the bridge makes in their immediate area. When abutments are part of retaining walls, texture and pattern can be effective in emphasizing the continuity with the wall.

Signing, Lighting, and Landscaping

Usually there are alternate locations for signs in interchanges. Every effort should be made to keep them off of ramp structures, as they will always seem to interrupt the visual continuity of the structure. If signs cannot be avoided, the size of signs for undercrossing roadways should be adjusted so that the vertical dimension stays below the parapet top. Sign structures on ramp bridges should be integrated with the superstructure design so the sign does not look like a bolted-on afterthought.

Sign structures are often part of viaducts. The long length of the structure makes them unavoidable, both for the undercrossing roadways and the viaduct itself. Again, the size of signs for undercrossing roadways should be adjusted so that the vertical dimension stays below the parapet top, and supports for sign structures on the viaduct should be integrated with the superstructure design. (See the "Signs on Bridges" section in Chapter Five.)

FIGURE 6-54 *Long spans and thin, widely spaced piers allow continued use of the space under this LRT line, preventing it from turning into neglected wasteland and restoring the connection between the uses on either side. JFK Airport Transit System, Long Island, New York.*

The lighting of the underside of urban interchange structures and viaducts is often important for reasons of function as well as appearance. Lights should be placed to coordinate with the major structural elements and the traffic patterns below. By taking this approach, the lighting can make a positive contribution to the visual impression, over and above merely providing light. Supports for roadway lighting should avoid interrupting the overall sweep and continuity of the structure. Widenings, or "blisters," tacked onto the parapet can be distracting unless detailed with a smooth taper. In interchange areas and along viaducts, it is sometimes possible to use higher, ground-mounted poles, which avoids the problems of mounting poles on the bridges.

Landscaping can provide an important visual contribution to both ramp and viaduct structures. An interchange landscaping plan can emphasize the major forms of the interchange and the continuity of the roadways while screening out distractions. Such a plan will make traversing the interchange both a safer and more pleasant experience. Landscaping and hardscaping areas under and near viaducts is an important way to knit back together the uses on either side, making the space under the viaduct a neighborhood asset rather than a neighborhood nuisance (Figures 6-54 and 6-55).

FIGURE 6-55 *Long spans allow park use of the ground plane under this viaduct for I-88 in Binghamton, New York. A more attractive or even more brightly painted superstructure and more attractive piers would have improved things still further.*

CASE STUDY

The Big I Interchange, High Level Ramps, Albuquerque, New Mexico[1]

The interchange where I-25 crosses I-40 in Albuquerque, New Mexico, was originally designed in the 1960s to handle 40,000 vehicles per day. By the late 1990s, the interchange was handling nearly 300,000 vehicles per day and was completely overloaded, as well as being physically worn out. Being in the middle of a major metropolitan area, the interchange could not be kept out of service for a long time. After discussing various closure options, the city and state decided to allow night-time closures for only a two-year period and completely reconstruct the facility within that ambitious time frame. A complete redesign of the interchange was undertaken, which increased the number of bridges in the interchange from 15 to 45. Eight of those bridges were high-level flyover ramps. The ramps had to be built while the interchange stayed in service beneath them. The physical constraints for the bridges included very limited opportunities for pier locations, the need to work over traffic, and to meet the tight time frame.

The visual environment is dominated by the roadways and graded slopes of the interchange itself. In keeping with the desert climate, it was anticipated that most of these slopes would be covered with desert vegetation or slope paving of crushed stone. The interchange is surrounded on all sides by the city's street grid, which interconnects with the interchange through a series of frontage roads. The nearby uses include low-level commercial and residential buildings and one mid-rise hotel. It is close to the downtown buildings, and from many locations appears to be part of the city's central core (Figure 6-56).

As the hub of vehicular travel in the Albuquerque region, most people in the region pass through the interchange regularly, many twice daily. It forms a big part of everyone's life and asserts a strong influence on the image of the city. Together with its frontage roads, the interchange's structures extend for long distances along the two interstate corridors, making it hard to designate a place where the interchange ends.

The design intention/vision by the city and state was to take this once-in-a-lifetime opportunity to create a civic asset for the community. The vision developed over a period of years. Studies of the urban design and aesthetic aspects of the interchange actually began with Project I, a non-profit corporation formed in the 1980s by civic-minded citizens whose goal was to create green space along New Mexico's highways. The organization worked with the City of Albuquerque and the New Mexico Department of Transportation (NMDOT) to develop a master plan for landscaping the interstate system in Albuquerque. This initial effort inspired a continuing focus on the aesthetics of all subsequent interstate reconstruction in Albuquerque. Influenced by Project I's goals, the city developed an Interstate Corridor Enhancement Plan in 1999 to define more specific design elements for the city's two interstate corridors. Its recommendations included design themes as well as methods of raising money for features not eligible for funding by NMDOT. The recommendations of these earlier studies became an integral part of all subsequent engineering studies of the interchange, and culminated in an aesthetic design plan. The design intention/vision was further influenced by the need to get the interchange rebuilt within the time available and keep traffic moving in the meantime.

The aesthetic planning efforts produced a formal aesthetic vision statement: "The vision for the New Big I is to create a unique freeway interchange that demonstrates a cohesive and

FIGURE 6-56 *Albuquerque, New Mexico's Big-I interchange brings together I-25 and I-40 close to the city's core.*

consistent quality of design through clean, simple, and timeless form that celebrates Albuquerque as the crossroads of the Southwest." Five specific goals were identified to define the vision statement:

- Design the Big I so that it is in harmony with the overall city environment.

- Incorporate as many concepts as possible from the previous planning documents.

- Develop visual enhancements that are complementary to the surrounding physical environment.

- Create a unique visual experience not only for visitors to Albuquerque, but also for local residents and commuters.

- Develop aesthetic enhancements that ensure the safety of the traveling public.

The design team identified four design principles that drove the design:

- *Interstate travelers.* Traveling through the interstate takes only a few minutes. To be memorable, images need to be bold, clean, simple, and can register in the mind in a very short length of time.

- *Local travelers.* Local residents and commuters will experience the visual aspects of the project on a daily basis. Any design elements need to be pleasing to the local community.

- *Roadway in harmony with the city.* The Big I interchange is located in the heart of Albuquerque, a bustling metropolitan setting. The design elements should be complementary to the urban images that are present, and also to the southwestern historical imagery found in Albuquerque.

- *Views and vistas.* While the interchange is located in an urban setting, with the high-level freeway ramps within the interchange, travelers will experience views out to and beyond the interchange. The aesthetic design must take into account these new visual opportunities.

Three visual design concepts were identified to help achieve the aesthetic goals and principles:

- *Form and line.* The form of the bridge structures should be simple, and lines should be fluid so that they do not draw attention to themselves but support the overall, uncluttered composition of the bridges.

- *Color.* Color can also provide interest to or highlight specific elements. A special emphasis can be given to the glorious red sunsets that occur in the Southwest.

- *Pattern and texture.* Texture provides visual relief to an object, and bridge parapets are ideal locations to apply pattern or texture due to their elevated position.

In order to meet the fast-track schedule for the project, a task-force style of management was chosen for the project design. Given the importance attached to aesthetics, an aesthetics task force was created, led by an aesthetics design manager. This manager, who was given a level of authority equal to the other design task force managers, participated in all decision making about all features of the interchange from the beginning of the design process. The aesthetics task force included representatives from NMDOT, the City of Albuquerque, and major stakeholders, as well as the design team. The presence of NMDOT and the city injected fiscal responsibility into the concepts. The role of the task force was to ensure follow-through on the commitments of the previous studies, to present the aesthetic concepts to stakeholders and the public, and to facilitate the prioritizing of aesthetic concepts to meet funding limitations. The costs of proposed aesthetic treatments were estimated as they were developed and compared to the available budget, which permitted if necessary timely modification of the concept, its delay to a future phase, or its assignment to another implementing agency. This method was used for all aspects of the interchange design to ensure that the project met its budget. The creation and empowerment of the aesthetic task force was crucial to the aesthetic success of the project.

To provide early public participation, the aesthetic design team presented proposed treatments at two public focus meetings that were held about three months into the project design. Sketches of several different treatments were presented at the first meeting. Following a general comment from the community to "keep it simple," the team presented more refined concepts at the second meeting, which received general approval. The final concepts reflected the comments at the second meeting.

With 45 new bridge and 10 bridge widenings/rehabilitations, the consistent application of aesthetic concepts was critical to the visual quality of the completed interchange. Because of their length and cost,

FIGURE 6-57 *The structures for the Big-I interchange in Albuquerque permit open vistas through the interchange, with plenty of space for undercrossing roadways. The rosy tan color and the blue parapet stripe tie together every element of the Big-I interchange (see color insert).*

as well as their visual prominence, priority was given to the design of the eight high-level flyover ramp bridges at the center of the interchange. The geometry of the ramps was adjusted to keep the flyover ramps at two constant widths, one carrying one lane with shoulders and the other carrying two. Concrete box girders were chosen for the ramps, rather than curved steel–plate girders. Both box sections were single–cell shapes with wide overhangs, providing both ample shadow lines and a smooth and uncluttered soffit. The box girders are slightly trapezoidal in shape, which narrows the bottom flange and makes the boxes seem smaller and lighter. The parapets are uniform throughout and have a continuous inset strip with a simple vertical groove texture. The effect is to reduce the perceived depth of the structure. There is no railing and no pedestrian screen.

There were several reasons for this choice: The concentration of reactions at the centers of the boxes kept the pier columns thin, so that they did not require much space and were easy to place among the interweaving roadways; this also eliminated the potential clearance problems of hammerhead pier caps. That, plus the long spans available with concrete box girders, gave much flexibility in the placement of piers. All of these characteristics dictated that concrete box girders would do the best job of providing the simple, clean, fluid lines desired by the city and the aesthetic task force. Also, the girders could be manufactured locally, freeing the schedule from dependence on distant fabricators, and erected with minimal interference to traffic using the balanced cantilever method. Finally, the costs were comparable with those for steel.

Piers were placed to keep them as clear as possible of undercrossing roadways, mainly for construction reasons. The result was to provide generous vistas through the interchange (Figure 6-57).

The tallest flyover bridges soar 100 feet over the lowest frontage road. Because of their prominence, the design of these piers required careful attention. The piers comprise an octagon in plan, giving them a faceted appearance, which makes them seem thinner and taller. They flare outward at the top to accommodate the bearings and to visually blend them into the trapezoidal box girders. For economy, designers used a single-column section and a single flare throughout the flyover ramp structures, regardless of the width of the box girder (Figure 6-58).

Most of the lower-level bridges are precast I-girders on conventional multicolumn bents. However, the columns were detailed with an octagonal section and flared top similar to the flyover piers, and the same parapet design was used to establish visual continuity with the flyover structures.

Abutments were placed to minimize wall lengths and heights. Most abutments are small, and simply

FIGURE 6-58 *A single pier-top accommodated both one-lane and two-lane ramps (see color insert).*

facilitate the transition to embankment. Some tie into retaining walls. The parapet detail continues from the bridges onto the abutments and ends at a standard transition detail. The intent was to make the bridges seem longer. A modified parapet detail continues from the transition element across both types of abutments and forms a continuous coping along all adjoining retaining walls. The walls are all made of mechanically stabilized earth (MSE) panels. All panels were faced with a texture that is an abstraction of the ancient human stonework seen in nearby Chaco Canyon.

It is with color, however, that the Big I makes its major statement. All of the surfaces of all of the bridges, piers, parapets, roadway traffic barriers, and sign and light supports are coated with the same rosy tan color. Even the gravel slope paving shares the hue. The only contrast color is on the inset strips of the bridge parapets, which are coated with a bright blue (Figure 6-58). The rock-patterned retaining walls are all coated in a compatible tan color.

The rosy tan color ties together all of the disparate elements of the interchange and simplifies the view. Thanks to the color, the interchange seems to fit right into the Sandia Mountains and nearby developments, as if it had always been there (Figure 6-59). The blue stripe acts even more strongly to tie the whole complex together. At the same time, when seen against the sky, it seems to unite with the blue background, fuzzing the upper edge of the structures and making them seem thinner.

The bold use of color was seen as a desirable feature of the project going back to the original Project I report. Due to the importance of the project within the city, NMDOT decided to assign the color choice to the public, an action believed to be a

FIGURE 6-59 *Because of its color, the Big-I interchange seems to fit right into the Sandia Mountains and nearby development, as if it had always been there (see color insert).*

national first. Two color schemes were placed on a number of public Web sites for a two-week voting period. The winning scheme, by far the brighter of the two, received almost twice as many votes.

The box girders have no visible bracing and, with a few inconspicuous exceptions, there is no conduit or other piping to interrupt the curved surfaces of the bridges. The sign structures are simple, single curved tubes, carefully placed to keep them off of the structures. The lighting is ground-mounted high-mast lighting, so no poles appear on the bridges.

The Big I has clearly met the design intention/vision. It is truly "a unique freeway interchange that demonstrates a cohesive and consistent quality of design through a clean, simple, timeless form that celebrates Albuquerque as the crossroads of the Southwest." Moreover, it was finished within the two-year period, while keeping traffic moving throughout, and met its budget projections. The project clearly refutes all those who say that aesthetic quality cannot be achieved on a tight budget or that a tight schedule dictates the use of the same old, tired standard designs.

The Big I bridge design team was led by the URS Corporation, assisted by subconsultants Chavez Grieves and Parsons Brinkerhoff with Alex Whitney of URS as the bridge design manager. Robert Peters, now of the Parsons Transportation Group, was the lead aesthetic advisor. Any great project requires a great client, and the contributions of NMDOT and the City of Albuquerque cannot be overstated.

MOVABLE BRIDGES

Picture books on bridges rarely show movable bridges. Why is this? Because few people find them attractive (though many bridge engineers would disagree). Movable bridges present a number of aesthetic problems (Figure 6-60). The counterweights and structural elements that move and support them are difficult to make attractive. The lift towers and bascule piers can become massive and detract from the proportions of the bridge. There is often a structural and visual discontinuity between the movable span and its approach span, a problem that can be exacerbated by massive piers. Finally, movable bridges have often been seen as industrial structures, with little attention given to appearance. All that said, there have been design-

FIGURE 6-60 *Former 17th Street Bridge, Fort Lauderdale, Florida. A typical movable bridge of the past, with features that give such bridges a bad name. The piers of the movable structure itself are massive and block views through the structure, while the approach piers are numerous and too close together, blocking views as well.*

ers who faced all of these challenges, yet succeeded in creating movable bridges that are assets to their communities.

Movable bridges can, in fact, have important aesthetic advantages in urbanized areas: The minimized structure height above the water and the shorter approach structure can make them less disruptive, physically as well as visually. Also, movable spans can be more structurally expressive than a fixed span.

Movable bridges come in a large range of sizes and types, and this section addresses some of the more general issues, without attempting to cover every possibility. This type of bridge represents a special case of bridges over rivers and tidal waterways, so the two sections should be read together.

Developing a Design Intention/Vision

Physical Requirements

Low roadway profiles combined with the presence of a navigational channel will be the primary reason for the movable bridge. Major considerations will be the width and the height of the clearance that must be provided when the bridge is in the open position. When setting the height and span of the movable bridge, the nature of the marine traffic will be important. For example, if there will be substantial recreational boat traffic, the closed height of the bridge may be set to avoid recreational traffic, so that openings are necessary only for commercial shipping.

The need for and location of the operator's control house is a very important consideration. Modern technology sometimes permits these bridges to be operated from a remote location, in which case there is no need for an operator's house. More often, a facility must exist in a location that meets the requirements of the bridge owner and the marine regulatory agencies. The location and height of this facility will be a function of sight distances to the highway and marine traffic, access to the operating mechanisms for monitoring and maintenance, and operator access, all of which need to be understood in advance.

These bridges are often called on to provide pedestrian and bicycle access, as well as access for fishing, all of which need to be anticipated from the beginning. Swing and lift bridges will have one movable span with the lift span flanked by high vertical towers. Bascule bridges typically have one or a pair of movable spans. Bridges across wider waterways will have approach spans that can be a much larger portion of the total bridge than the movable span. They will share the concerns outlined in the "River and Tidal Waterways" section.

Visual Environment

The visual environment is established, first, by the nature of the waterway. For very wide waterways, the water itself will be the dominant aspect of the visual environment. For narrower waterways, the range of shoreline uses will have more influence, whether industrial, commercial, residential, developed park, or natural. The visual effect of the uses should be confirmed in the field with photographs, since riverbank vegetation often screens from view the actual use. Movable bridges often occur in urbanized areas, and a review of nearby buildings is necessary to understand the full visual environment. The resulting information should be used to guide decisions related to the visual environment. For example, the towers of a lift bridge may seem out of place in a residential environment, and that visual impact needs to be understood.

Nearby and Associated Uses

Public shoreline open space, such as a riverfront park, fishing pier, or festival marketplace, will be very important in determining viewpoints and the nature of the design. Commercial or residential uses offering a number of potential overlooks will also be important considerations. Industrial uses offering no public access will require less consideration.

Symbolic Functions

Movable bridges typically have proportionally more massive piers, as well as control houses, lift towers, and visible lifting elements, such as counterweights and counterweight supports, all of which make them more prominent in the visual scene than most bridges. Their appearance changes dramatically when they are in the open position, and this also needs to be understood (Figure 6-61). Thus, they have more potential to become a landmark, whether asset or eyesore. The goal must be to do them well, so that they become a visual asset to their surroundings.

Boundaries

Because the structural requirements of the movable portion of the bridge are so different from the approaches, there is a danger of visually subdividing the bridge itself. The goal should be to develop a consistent concept that carries across the entire length of the project. Where changes in structure type are necessary, they should be marked by logical structural elements. Anything that can be seen at one time from the important viewpoints should be considered part of the structure for design purposes.

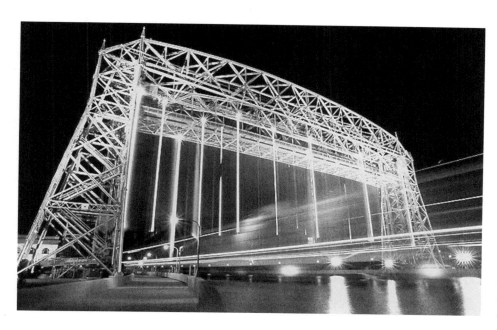

FIGURE 6-61 *Movable bridges are all about movement— marine, automobile, and the bridge itself—as this time exposure of the Duluth Aerial Bridge makes clear. (See also color insert.)*

Guidelines for the Design Intention/Vision

The most important views for movable bridges are usually oblique views from points along the shore. Particular locations can be identified that are likely to be the most-favored viewing points, such as public docks or parks, where people are likely to congregate. Since a movable bridge will likely be on a waterway with a significant amount of recreational boat traffic, the view from the water should also be considered. The point of contact with the shore can also be an important feature in forming the visual impression, particularly if there is pedestrian use or a shoreside roadway at that point. Photographs should be taken from the most important viewpoints and used as the base for future sketches and renderings.

At the most likely oblique viewpoints, the most important features in determining the visual impression will be the size and shape of the bascule, swing, or lift piers and any visible counterweights, along with the superstructure type or types, and the transitions between the movable span and the approaches. This means that movable bridges cannot be prioritized by element as easily as other types of bridges. The main span superstructure, pier placement, and pier shape must all be considered at once. If the abutments are within a span or two of the movable span, they also must be considered at the same time. The pier shapes on the approaches are the next most important consideration, particularly as seen at an oblique angle, with piers lining up one behind the other. Finally, parapet and railing design will be important because of their potential to provide a unifying element that extends across both the movable span and the approaches. They will also be very important features to users of any sidewalk or bike path.

In this type of bridge, the movable span will inevitably play the starring role. At the same time, bridges of this type can be quite long and are usually seen in their entirety from the most important viewpoints. Thus, the goal must be to bring the "supporting cast", the approaches and abutments, into a complementary role, so that the bridge comes together as a whole. There are two basic options: The first is to focus all attention on the movable span, with everything else kept as neutral as possible. This probably works best when the movable span is 50 percent or more of the bridge length. In the second, elements are developed that are shared between the movable span and the approaches, such as common girder shapes, pier shapes, or parapet and railing configurations, so as to unify the movable span and its approaches. When designing a movable bridge:

- Decide on the relationship of the movable span and the approaches and follow it consistently.

- Minimize the size or apparent size of the bascule, swing, or lift piers, and/or make them as transparent as possible.

- When large piers are unavoidable, integrate them into the design as a whole, coordinating with the approaches as well as the movable span.

- Integrate the operator's house into the overall concept or place it off the bridge.

- Keep the approach spans as long as possible to keep views open and minimize the "forest of piers" look.

- Keep approach pier shafts as thin as possible to make the structure transparent.

- Except for very short bridges in urban environments, emphasize the horizontality of railings and parapets; avoid pedestals, pylons, and pilasters that break the horizontal sweep of the structure.

Determinants of Appearance

Because the key parts of a movable bridge all have to be considered together, their Determinants of Appearance are presented in a different order than for other bridge types. However, the same basic rule applies: Determine the largest, most visible elements first.

Geometry

Longer bridges incorporating movable spans should follow the guidelines suggested for bridges over rivers and tidal waterways. Shorter bridges over rivers and canals will probably have their geometry determined by the approach road or street.

Movable Span Type

Many factors will determine the choice of type of bridge, bascule, swing or lift, many of which are outside the scope of this book. However, appearance should be one of the key factors to consider when selecting the type of movable span. Lift bridges are most problematic because the lifting towers constitute a permanent element in the scene, more visible in a flat landscape, less so when surrounded by tall buildings or industrial structures. When the decision is made to construct a lift bridge in a visually important environment, it must be accompanied by a commitment to make the towers visually attractive. Just like any other bridge, a lift bridge can become an asset to its neighborhood when it is well done. Sacramento's Tower Bridge is considered a symbol of the city. Even more dramatic is the Aerial Bridge in Duluth, Minnesota (Figure 6-61, page 251).

The visual impact of bascule bridges depends, first, on whether the counterweights and counterweight support arms are visible, and, second, on the apparent massiveness of the support pier. The first goal should be to place the counterweight so that it is out of sight in the closed position. If that is not possible, the counterweight and its supporting structure may become prominent elements in the scene. Just as with a lift bridge, when a decision is made to construct a bascule bridge with a visible counterweight mechanism in a visually important environment, it must be accompanied by a commitment to make the mechanism visually attractive. This can be done by making the counterweight shape and the shapes of its supporting elements simple and consistent with the shapes of the other elements of the bridge. If the counterweight and its mechanism can be hidden, the focus should be on minimizing the apparent size of the bascule pier and integrating it with the rest of the bridge.

The pivot and rest piers for swing bridges typically lie entirely below the superstructure. Thus, they don't interfere with the visual continuity of the superstructure, and their massiveness is literally overshadowed by the superstructure above. For this reason, swing bridges generally are the least visually intrusive movable bridges (Figure 6-62).

FIGURE 6-62 *This historic swing bridge exhibits principles that still apply today. The truss is deeper over the swing pier, reflecting the greater forces there. The outline of the bridge is refined into an attractive shape, and the truss diagonals all occur on the same horizontal module, making the bridge simpler and easier to understand.*

Pier Placement and Shape and Operator House Placement

The location of the lift towers or bascule or swing piers will be determined by the location of the navigational channel. Given the normally massive size of these elements relative to the rest of the bridge, the quality of their appearance will be critical to the appearance of the bridge. Designers of lift piers have taken variations on two basic approaches. Either the structure is entirely enclosed with a nonstructural skin, allowing the movable counterweights and its mechanisms to be hidden, or the whole thing is opened up, so that the sheaves, ropes, counterweights, and their supporting structure are all exposed. The first approach can give the structure a clean but very massive appearance. The second approach can work well, but requires uncommon care in the design of the structural and mechanical elements so that the whole ensemble is attractive (Figure 6-63).

Bascule piers are commonly very massive elements. In the past, these piers were often conceived as single forms extending from water level to the top of the operator's house, and sized to contain the required footprint of the operating mechanism and the operator's house. In the higher bascule bridges now seen more frequently, designers realized that the machinery and operator were occupying only a small frac-

FIGURE 6-63 *The machinery that makes a bridge move can be fascinating and, if well designed, attractive.*

tion of the total volume. Newer designs have tucked them into enclosures within and between the structural piers and girders. The result has been much more open piers and much more transparent bridges, such as the 17th Street Causeway discussed in the case study at the end of this section (Figure 6-68, page 262).

For lower-level bascule bridges, designers must find some way to minimize the apparent size of the bascule pier issue, such as to:

- Visually divide large piers into smaller elements using elements such as a pier overlook or stepped pier face.

- Use strategically located windows or louvers to visually divide the pier and provide ventilation and access to machinery and electrical equipment.

- Divide the piers into more than one element. For example, rolling bascule piers are often actually a forward pier and a rear pier with a track girder between them.

Swing piers generally are the least massive in appearance relative to the rest of the bridge. The design goal should be to keep the pier simple and consistent in shape with the other elements of the bridge.

For the approaches, the number of piers becomes a significant visual parameter. The fewer the piers, the less likely they will appear to line up behind each other; hence, the bridge will appear less wall-like when seen in the oblique view. Span length should be as great as possible within the constraints of affordability. On the approaches, designers should take their cue for pier shapes from the shape of the bascule, swing, or lift piers so as to create an appearance of continuity across the whole bridge. Designers should also follow the guidelines described in the "Rivers and Tidal Waterways" section.

Lift towers or bascule or swing piers will be in close relation to the navigation channel and the operator's house can be made an extension of them (Figure 6-64). However, the operator's house need not necessarily be tied to the same place. Using technology for remote observation or just taking advantage of available sight lines, it may be possible to physically separate the operator's house from these elements, reducing their size and consequent design problems. On wide modern bridges, there is sometimes sufficient median width to allow placement of the operator's house at roadway level in the median, where it appears separate from the supporting elements.

FIGURE 6-64 *The operator's house is a transparent extension of the bascule pier of this bridge in Quogue, Massachusetts.*

Superstructure Type

Movable structures are usually girders or trusses, though movable cable-supported and arch pedestrian bridges have recently been built. Steel box girders are a good alternative to trusses or plate girders where a clean uncluttered look is desired. Their stiffening and bracing elements are all inside and hidden from view. Box girders also have a more similar look to concrete structures than plate girders or trusses, which can make them more compatible with a box girder approach. It is critical to make a clear and simple transition between the movable span and the approach spans. For example, a well-designed bascule pier can be used to distinguish the movable span from approach spans while maintaining visual continuity through detailing. The bascule pier, handrails, light poles and fixtures, colors, overhangs, and other features can be the same or similar to those on the approach spans and abutments. Many examples exist of truss type bascule bridges with arched approaches that are quite well balanced visually due to the similarities between the massive bascule piers and massive abutments, even though the structure type of the approaches and movable span are very different.

Abutment Placement

Similar to any water crossing, the shoreline is the obvious place for the abutments (see the guidelines in the "Rivers and Tidal Waterways" section). If one of the movable span piers is close to the shore anyway, the placement of the abutment should be considered relative to the lift pier. It may be possible to place the abutment such that an intermediate pier is avoided, thus simplifying the appearance of the structure. If the navigation channel is along the shoreline, the abutment becomes the bascule, swing, or lift pier, and all of the preceding discussion applies.

Superstructure Shape

Bascule and swing bridges are cantilever bridges, and look best when the girder/truss is haunched. The tapered movable element can gracefully accommodate the change in depth from movable span to approach spans. The haunches visually punctuate the sweep of the bridge and draw attention to the movable span. Both the closed and open positions of the movable span must be studied visually. For bascule and swing bridges, the visual continuity of the bridge is broken when the span is open. For vertical lift bridges, the appearance with the bridge open can be quite

FIGURE 6-65 *Best to keep small movable bridges as simple as possible, a strategy supported in this Annapolis, Maryland, bridge by placing the operator's house on the shore.*

intimidating. For bascule bridges, the angle of opening of the bridge must be considered (some open only 55 to 60 degrees, while others open almost 90 degrees).

There is a debate among designers about deck overhangs. Structural efficiency points in the direction of relatively wide overhangs, but operational stability encourages narrower overhangs. Aesthetically, wide overhangs generally look better and permit narrower piers with fewer columns, producing a more transparent bridge. In any case, keeping similar deck overhangs over both the movable span and the approaches will create a constant shadow line that will emphasize the visual continuity of the bridge.

The parapet and any railings should be used to reinforce the horizontal continuity of the superstructure. Because small details will not register on bridges of this size, the design should use large offsets, insets, or repetitive patterns that emphasize horizontal continuity. See the railing design on the bridge in Figure 5-56 (page 184), which emphasizes the horizontality of the structure.

Operator's House Shape

The operator's house is a piece of architecture, in that it provides a space for the operator and the maintenance staff; it has windows, doors, and all of the other necessary features of a building. But it is also part of the bridge, which is usually by far the larger and more dominant element. A frequent practice is to treat it as a piece of architecture and design it in relation to nearby buildings. That sometimes results in the details of the operator's house being extended to the piers, parapets, and railings of the bridge, so that the operator's house ends up being the most important element of the design. This is actually a reasonable strategy for small bridges in urbanized environments, particularly in the presence of a strong architectural or historic environment that must be respected. In contrast, for large bridges, the operator's house is too small a part of the overall bridge to carry that visual load. Better in these cases to develop the bridge design based on the optimum shapes of structural elements themselves, then integrate the operator's house by incorporating forms suggested by the structural elements of the bridge.

Abutment Shape

If the abutment is the lift pier, then all of the earlier discussion about lift piers applies. If there are only one or two spans between the bascule pier and the abutment, the abutment becomes a key element of the bridge and must be designed as a related element of the bascule pier. The abutment and bascule pier may flank either side of a pedestrian walkway or even a roadway that runs along the edge of the waterway. In both cases, the presence of related uses on the shore, such as walkways, roadways, and adjoining buildings, needs to be considered in the design. In cases where the abutment is some distance from the movable span, its small relative size compared to the structure as a whole means that its shape will have little effect on the overall appearance of the bridge. However, its design must still recognize shore-side uses.

Color

Color should be used to integrate the movable span with the rest of the structure. Similar elements, such as approach girders and bascule girders, should be painted the same color across the whole structure. Where steel and concrete girders are used together, the steel paint color should be selected to match the concrete. Bright color applied to a secondary element, such as a railing, can be used to visually pull together different structural types.

When selecting colors, the apparent color of the water will be an important consideration over the wider waterways. Apparent water color is a function of the color of the water itself, which can vary by season depending on runoff and temperature, and sky color, which will also vary by day and season. In urban areas, the background of nearby buildings and traditional colors of the area need to be considered.

Surface Patterns, Texture, Ornamentation, and Details

For the same reason as for bridges over rivers and tidal waterways, traditional brick and stone are generally not convincing when used as superstructure components of these bridges. However, in an urbanized area, they might be considered as part of the design of the operator's house and/or for pier and abutment facing. Such materials should be used in ways consistent with their traditional uses, which is as a load-bearing material. Thus, they should be carried down to a concrete base or the water-line. They should not be supported by the superstructure and certainly not on the movable span.

Most of the bridge will be in view over the water at one time from many viewpoints. A movable span will have visual points of interest resulting from the operating mechanism and the structure itself. Drawing attention to these elements by adding color or detail can improve the appearance of these structures. If there is a pedestrian walkway on the structure, overlooks at key piers will punctuate the sweep of the structure and provide a pedestrian amenity.

Pedestrian walkways on river crossings must be made wide enough to be comfortable for two people to walk abreast or pass one another: 8 feet should be the minimum, with 10 feet the minimum for walkways that are expected to get significant use or bicycle traffic. Patterns in the walkway surface created by scoring, color in the concrete, or changes in materials should be used to visually break down the apparent length of the journey into manageable increments.

Signing, Lighting, and Landscaping

Movable spans are often called on to support sign and signal structures. Integrating these sign structure with the superstructure and/or parapet/railing design is the important factor. It is important to place them in some logical relation to the overall bridge, such as at a pier, and to smoothly integrate the connection of the sign structure into the overall structure. The sign structure should seem like part of the original concept, not a bolted-on afterthought.

The size, spacing, and style of lighting fixtures may have a significant impact on appearance, and should be considered in relation to the overall design intent for the bridge. The fixtures will produce reflections in the water that will magnify their impact at night. Sharp cutoff fixtures can reduce the over-spill of light. Pole spacing should coincide with, or be an even multiple of, span or the spacing of other structural features. In addition, if the lighting is to continue along an adjoining roadway or river walk, a light fixture and pole spacing should be selected that fits the overall concept and detailing of the bridge.

Movable bridges in prominent locations can be very effectively lit at night to create a nighttime landmark. The reflections multiply the effect. The most effective approach is to light the piers and/or underside of the structure, which takes best advantage of the reflections. This is a difficult task, one well worth the investment in the services of a specialist.

Most movable bridges are required to include barrier gates to prevent traffic from getting too close to the roadway gap resulting from the opening of the bridge. Mostly, these gates and barriers are of the semaphore type, so they are always vertical and exposed from many viewpoints. These features and their accompanying traffic signal supports need to be given careful consideration in design consistent with their prominence.

Landscaping elements are too small to have much of an impact on the appearance of the overall bridge. However, landscaping may be a very important part of shoreline uses, and may affect the appearance of bridge elements on shore, particularly abutments.

CASE STUDY

17th Street Causeway Bridge, Fort Lauderdale, Florida

In Fort Lauderdale, 17th Street is a major connection between the Atlantic beaches and the center of the city. The bridge is located adjacent to Port Everglades, a major East Coast port facility and numerous other marinas (Figure 6-66). Fort Lauderdale is a highly popular yachting destination, and many boats traverse the intracoastal waterway (Stranahan River) through the bridge. Many are sailboats, some with mast heights approaching 130 feet. The existing bascule bridge was a four-lane structure built in 1956 to a

FIGURE 6-66 *The overall view of the new 17th Street Bridge, Fort Lauderdale, Florida, shows the space made available for shore uses under the approach spans.*

standard Florida design with an exposed counterweight, a massive bascule pier, and short approach spans supported by concrete pier bents (Figure 6-60, page 249). The vertical clearance when closed was only 25 feet, so that almost every boat passage required an opening. Automobile traffic has also rapidly increased in recent years, and the resulting conflicts were creating gridlock on both 17th Street and the Stranahan River.

The physical requirements were to provide a vertical clearance of at least 130 feet and a horizontal channel width between fenders of 125 feet. The visual environment is dominated by boats themselves, their wharves, and the waterfront activities associated with the port and the marinas. Nearby uses consist almost entirely of low-rise commercial buildings that serve the port and marinas. The bridge is considered a gateway to the famous beaches of Fort Lauderdale, so the community wanted a "signature bascule bridge" at that location. The physical boundaries of the project are the points at which the approaches return to the grade of the existing street system, but the visual boundaries as seen from the important river views are essentially set by the points where the approaches disappear between flanking buildings.

During preliminary studies, a wide range of options were considered as solutions to the problem. A tunnel was dismissed because of cost. A high-level (130-foot) bridge was dismissed because its lengthy approaches would be too disruptive to the highly urbanized areas on either end of the bridge. A lift bridge was dismissed because its 130-foot-plus high towers would have greatly exceeded the height of nearby buildings. A swing bridge was not practical because it would have required too much of the already limited vessel queuing space. A single-leaf bascule clear channel width would have required too long a span. The final conclusion was a twin-leaf bascule with a total span of about 230 feet and a vertical clearance when closed of 55 feet. This increase in vertical clearance has materially reduced the number of openings.

The Florida Department of Transportation (FDOT) takes a three-tiered approach to determining the aesthetic importance of their bridges. The 17th Street Causeway was designated as Level Three, which entitled it to the highest degree of aesthetic consideration. In addition to opening up the possibility of innovative and well-shaped structural members, Level Three also allows for landscaping and collateral development to support the adjoining community. Two "design charettes" were held to solicit the views of the community. Participants included concerned citizens, local government staff and elected officials, FDOT staff, and the design team. The first charette focused on familiarizing the community with the design issues and possibilities of the site. The goal was to gain agreement on design criteria that could form a statement of design intention/vision for the project. The resulting list of criteria can be summarized as follows:

- Thin, elegant, and simple contemporary structure

- Graceful, clean lines; free of adornment

- Compatible with the environment

- Each element important to the whole

- Railings that allow outward views

- White or gray concrete

For the second charette, the design team did some conceptual engineering, looking for ways to meet the community's goals, as well as the functional goals of the project. The team began with an investigation of ways to reduce the massiveness of the bascule piers. This was particularly important as the piers would be roughly two and one-half times as high as the existing piers. One approach (Figure 6-67.1) focused on

FIGURE 6-67.1 *Drawings presented to the community for the Narrow Pier option.*

making the pier as a whole as narrow as possible in the longitudinal direction while splitting it into five elements in the transverse directions to permit longitudinal views through the pier. The key decision was to place the counterweight outside the bascule pier, which dramatically reduced the required volume of the pier. In the closed position, the counterweight would lie between the approach girders and thus be hidden; in the open position, it would be visible. Though this pier would have been a marked aesthetic improvement over a typical bascule pier, it suffered functionally from restricted space for the machinery and live load bearings. The designers also felt that it was not enough of a departure to meet the goal for a signature bascule bridge.

The second approach (Figure 6-67.2) was in part generated by a search for structural and visual continuity between the bascule span and the approaches. The pier is V-shaped in the longitudinal direction, which means the pier is wide at the top where space is needed for the counterweight, machinery, and live load bearings, and narrow at the bottom, opening up views through the bridge. The longitudinal width at the bottom is only 7.5 feet, about one-third of the width of the first option. The face of the V was also opened up, making the pier even more transparent. In the transverse direction, the pier was divided into three elements, which was possible because the approach span is integral to the pier and does not require a separate support. The resulting openings allowed further views through the structure. The shape of the pier also allowed the trunnion to be placed about 10 feet closer to the channel than with the previous option, reducing the overall span and the weight of all of the movable elements. The pier became known as the *Carina* pier, which is the Latin word for the delta-shaped keel of a boat. The faces of the pier were flared out laterally to accommodate the pedestrian overlooks within the overall profile of the pier. The distinctive and unique shape of the piers makes the bridge truly a signature bascule bridge (Figure 6-68).

A tapered steel box girder section was chosen for the bascule superstructure because of its high torsional rigidity and reduced depth. With a box girder, all stiffening and bracing elements are inside the girder. The smooth outer face is easier to maintain and presents a clean and uncluttered appearance, which

FIGURE 6-67.2 *Drawings presented to the community for the Carina Pier option.*

is particularly important for a bascule span because when the bridge is open, the underside of the leaf is its most visible element. The combination of box girders and an exodermic concrete deck also eliminated the need for lateral bracing, further simplifying the appearance of the structure. The angle of rotation was chosen so that in the open position the bridge offers unlimited vertical clearance between the fenders (Figure 6-70).

The same criteria—simple, elegant, and contemporary—governed the design of the operator's house (Figure 6-68). The house reflects the curved surface of the piers, but in stainless steel. It is placed in the

FIGURE 6-68 *Because the machinery and counterweight is near roadway level, the bascule pier itself can be open and transparent. The operator's house reflects the smooth, curved shapes of the bascule pier.*

FIGURE 6-69 *The approach girders of the 17th Street Causeway Bridge in Fort Lauderdale, Florida, blend in to the bascule pier. The approach spans are long enough to allow views through the structure and to allow recreational uses of the area beneath the bridge.*

median and is seen as a separate element from the piers so that the perceived mass of the piers is minimized. There are four lines of railings. The lines between the roadways and the sidewalks are of an open, crash-tested design consisting of two horizontal rectangular tubes bolted to I-beam posts. They shield the pedestrian railing on the outside of the bridge (and the pedestrians) from vehicular impact, allowing more flexibility in the design of the pedestrian railing. The pedestrian railing consists of round aluminum posts curved inward at the top with a top hand-rail of large-diameter tube and horizontal small-diameter tubes below. The railing provides security for pedestrians while opening up the outward views to both drivers and pedestrians.

Large portions of the approaches are over parking and pedestrian activities associated with the neighboring uses. Multiple I-beam alternatives were considered but rejected primarily because of the dark, cluttered underside they would have presented to the users below. Instead, concrete box girders were selected because their wide overhangs and smooth reflective surfaces will provide a bright, open appearance, even from directly below. Aesthetic criteria also affected the design of substructure elements. Footings will be submerged below the mudline in order to minimize the visual impact of the supporting structures, as well as reduce the hazard to boating. The piers flare at the top to match the angle of the trapezoidal box section (Figure 6-71).

The Carina piers provide a unique and attractive shape that is an appropriate centerpiece of its marine environment. The approach spans encourage pedestrian and parking uses under the bridge. The operator's house and smaller elements such as the railings and lighting

FIGURE 6-70 *The position of the operator's house allows monitoring of all marine and vehicular movements regardless of the position of the bascule leaves.*

FIGURE 6-71 *Lighting brings out the shape of the Carina Pier at night. (See also color insert.)*

provide points of focus and continuity. The bridge overall forms a unified statement that meets the community's design intention/vision, giving citizens the signature bascule bridge they requested.

The design team for the bridge included E.C. Driver Associates (now part of URS Corporation) and Figg Engineering. Driver had primary responsibility for the movable span, and Figg had primary responsibility for the approach spans. The project manager for Driver was James M. Phillips III, and for Figg, during the concept development, John Corven.

RAILROAD BRIDGES

This section concerns bridges carrying railroads. Bridges carrying highways over railroads are similar in many respects to highway-over-highway bridges and viaducts. The ideas for such bridges can be derived from those sections of the chapter. Bridges for transit lines represent a different category. The loads imposed by transit trains or cars, whether "heavy" rail or "light" rail, are substantially less than those imposed by railroads. The structural types and dimensions are similar to those used for highway construction. The visual guidelines developed in the other sections of this chapter apply to transit structures as well, particularly the section on ramps and viaducts.

Developing a Design Intention/Vision

Physical Requirements

The railroad's alignment is typically less flexible than a highway's. The clearance requirements imposed by the railroad will be a controlling factor. If the structure is part of an existing railroad, the need to keep the track(s) in service will impose

FIGURE 6-72 *The deck overhang helps this structure seem thin; the abutments are set back so that views through the structure are opened up.*

construction-staging requirements that will influence the type of structure and the methods of construction. On streets and arterials, required provisions for pedestrians often create both design problems and opportunities.

Visual Environment

Railroads often abut industrial areas, and bridges there will be seen as just a part of a larger industrial complex. In contrast, grade separations are required in residential, commercial, agricultural, and wooded areas, where the structure can be the largest feature of the street scene or neighborhood of which it is a part, hence will attract the attention of people in the area. Its position as a visual focus needs to be recognized and understood.

Nearby and Associated Uses

If there are nearby residential and commercial buildings, they will become the backdrop for the structure, thus it will be necessary to consider the compatibility of the structure with the nearby buildings. If there are to be pedestrian uses through or around the structure, their needs should be considered. The emphasis must be on clear visibility to all parts of the pedestrian walkway and a clean and pleasant environment under the structure. An often-overlooked criterion is the need to provide clear lines of sight through the structure. Railroads can literally and figuratively divide communities, presenting a "Chinese wall" appearance. Extensive visibility through the structure along the connecting street can help to visually knit the community back together.

Symbolic Functions

As utilitarian facilities, the symbolic functions of railroad bridges are usually limited. However, the position of a particular structure in a particular streetscape may give it symbolic importance. For example, one could imagine a grade separation project on the main street of a small town that could become the symbol of reuniting a town formerly divided by an at-grade railroad line. It would be important to design the structure so that the functional division of the at-grade tracks is not replaced by an equally disruptive visual barrier. Railroad passenger stations are not the entry points they once were; many have become the centerpieces of historical redevelopment areas. A bridge in the vicinity could become an important symbolic adjunct to their historical revival.

Boundaries

Along the line of the railroad, the boundaries of the project will most likely be the abutments of the structure. Along the line of the undercrossing street, the boundary will depend on the extent of any retaining walls that are part of the project, as well as the extent to which the project is part of a larger streetscape or pedestrian environment. Then it will be important to continue the wall treatment, parapets and railings, paving, lighting, and other elements far enough to reestablish the visual as well as the physical connection through the structure.

Guidelines for Developing a Design Intention/Vision

The most important viewpoints for railroad bridges are typically associated with the undercrossing roadway. If that roadway is part of a freeway system or an arterial that carries only highway traffic, the important views of the structure will be no closer than 300 to 500 feet. At any point closer, viewers will be looking through the structure to features beyond. The time available to appreciate the structure will be similarly limited, depending on the point at which it first becomes visible and the speed of approach. For such structures, the important visual features will be the large elements. Details smaller than about 4 inches will not be noticed. Structures over arterial streets and roadways with pedestrian traffic can be seen at a leisurely pace and at fingertip distance. Decisions about details and materials will have greater impact on the visual impression.

Bridges carrying railroads create an entirely different visual impression than the typical highway structure. Because of the heavier railroad loads and the reluctance of railroads to use continuous structures, the superstructures become very deep and appear heavy. The most important features visually are likely to be the superstructure type and shape, especially the depth-to-span ratio; next will be pier and abutment shape, including the bearings; and finally will be superstructure details such as stiffeners.

While there are some design tactics that can be used to give the structure additional grace, the inherent strength and weight of the structure must be recognized. The best impression is created when this strength is displayed, while at the same time the apparent length of the structure is maximized. Making these deep structures appear graceful presents a real challenge to the designer. Unfortunately, the design of ordinary railroad bridges moves in just the opposite direction. Piers are extended vertically to hide the bearings, which also visually cuts the girder into shorter lengths and makes it appear deeper, while the vertical stiffeners visually shorten and deepen the girder even more. When designing a railroad bridge:

- Try to reduce the apparent depth and weight of the structure and extend its apparent length.

- Avoid the use of pilasters and pylons that subdivide the structure and shorten its apparent length.

- Expose the bearings so that the girder rests clear of the substructure and looks lighter.

Determinants of Appearance

Geometry

There is rarely much room to maneuver with the geometry of the railroad. Any flexibility that may be available will be in the alignment of the undercrossing street. That should be checked to see if it is possible to adjust the geometry to enlarge views through the structure or to create a more graceful transition with the roadway on either side of the bridge.

Superstructure Type

The most common type of structure for railroad bridges is a welded-steel-plate girder. In recent years, there has been some use of post-tensioned concrete and steel-box girders as well. Truss bridges were frequent decades ago, but are now rarely used except for spans whose discussion is beyond the scope of this book. The decision is generally controlled by economics and the preferences of the railroad's engineering staff. Concrete may have some visual advantages because there is no need for stiffeners, and the color is lighter than A588 steel. But these differences are not compelling.

It is important not to interrupt the view of the girders with pilasters or cheek walls at the piers and abutments. If not interrupted, the girder will seem to be one horizontal unit with an apparent depth-to-span ratio for the whole unit that is thinner than the actual ratio for each individual girder (Figure 6-73).

Arches have been used recently in urban situations where the goal was to create a more transparent structure that was more inviting to pedestrians (see the case study at the end of this section).

FIGURE 6-73 *This attractive railroad bridge is not matched by abutment walls of equal quality; a wall with a simpler surface that extended the line of the deck overhang and extended horizontally with the railing would have integrated better with the bridge.*

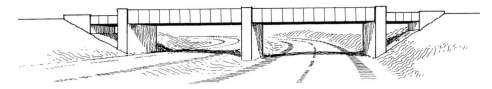

FIGURE 6-74.1 *When looking at an elevation drawing, the addition of side-spans would appear to open up the structure; in reality, the massive shoulder piers block the views through the sidespans. The change in girder depth also breaks up the lines of the structure, making the main span seem shorter and deeper.*

Pier Placement and Abutment Placement

Because of the loads there is a natural desire to keep the spans as short as reasonable. For spans over one or two roadways, this can mean simply placing piers at the points permitted by the undercrossing roadways. That said, the spacing should appear logical and not produce variations in girder depth. Keeping the girder depth the same over the entire structure will make the bridge seem longer and therefore thinner. Short spans over side slopes are a particular problem. If the girder is kept the same depth over the smaller span, the short-span girder will look unreasonably heavy. If the short-span girder is made thinner, a jump in girder profile will be created. In any case, little is gained visually by the additional spans, as the heavy shoulder piers block most of the additional view through the structure (Figure 6-74.1). It is better to leave out the end spans and use the money instead to place the abutments as far from the edge of the shoulder as the budget will allow (Figure 6-74.2).

Superstructure Shape

Railroad bridges present limited opportunities for influencing appearance, but those available should be explored. Girder dimensions and details can be used to emphasize apparent thinness and horizontality, and structural depth can be kept constant or smoothly varied over the entire bridge. For deck girder bridges, a limited overhang may be available, which can be enhanced to establish a horizontal shadow line that will make the bridge seem thinner.

When it comes to shape, railroad bridge girders can be analogized to Henry Ford's Model T. Model Ts came in one color: black. Railroad bridge girders come in one shape: rectangular. One variation available for a through-girder is to round the upper corner of the girders. This is a worthwhile, if subtle, detail because it adds a touch of interest that would not be there if the girder were just sliced off. Tapered or haunched girders are worth exploring, too, if the railroad will permit continuous structures, or to provide visual continuity between long- and short-span girders.

Details of the stiffeners and bearings are the primary areas for making the girder appear thinner and more graceful. Use high bearings, pedestals, or chamfers at pier tops to visually attenuate the bearing point. High-bearing shoes with pin connections to the girders, in particular, can exaggerate the apparent length of the girder (Figure 6-75). They also make the girder appear lighter because the girder is supported by a feature that seems very small compared to the size of the girder. In order to take advantage of this possibility, the shoes must be visible. In no case should

FIGURE 6-74.2 *Placing the abutments back on the slope opens up the structure more effectively than adding sidespans.*

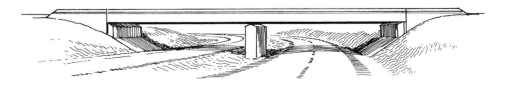

FIGURE 6-75 *This through-girder bridge shows the advantages of exposed high bearings and variable stiffener spacing in making a bridge seem longer and thinner. By holding the constant-depth girders up off the pier caps, the bearings allow the horizontal lines to be seen, making the bridge seem longer and thinner. The signs over the opposing lanes disrupt this effect and would have been better placed elsewhere.*

pilasters be used to hide the bearing. Horizontal stiffeners can be used to visually subdivide the girder and make it appear thinner. This is another application of the visual illusion illustrated in Figure 2-23 (page 42). Vertical stiffeners can help if their spacing is allowed to vary according to the shear stress. This creates a pattern that varies horizontally (the stiffeners are closer together over the supports), encourages our eyes to move horizontally, and, therefore, exaggerates the horizontal dimension of the girder. Deck girders require a parapet and/or railing. The details of these features should be developed to enhance the horizontality of the structure.

Pier Shape

Notwithstanding the techniques just described, the girder will still seem large and heavy. It is necessary that the pier seem to be clearly capable of supporting this weight. Any details should seem to reinforce the pier's weight-carrying capability. The emphasis should be on large, simple shapes. By its shape the pier can enhance the apparent length of the girder. A pronounced chamfer at the top of the pier will prevent the lines of the pier from visually interfering with the lines of the superstructure so that the superstructure appears longer and more continuous. It also reinforces the visual effect of high bearings.

Abutment Shape

The preceding comments about making clear the pier's capability to support the weight of the girder and using details that enhance the apparent length of the girder apply to the abutment as well. Other guidelines are to batter the face of the abutment, which reduces the apparent height of the abutment and makes it appear to have a greater carrying capacity, since it appears wider on the bottom. Wing walls should be aligned parallel to the railroad, and parapet and railings should be carried out to the ends of the wing walls. Bearings are a major structural element and should be left exposed. Taking these steps will have the effect of making the bridge seem longer and, therefore, thinner.

Color

Color can provide an inexpensive way to reduce a girder's apparent depth by dividing the girder horizontally into different areas of color. It helps when a hori-

FIGURE 6-76 *The apparent depth of this railroad bridge is reduced by the band of contrasting color above the horizontal stiffener (see color insert).*

zontal stiffener forms the line of division because it provides a visually logical location for the change, as on the bridge in Figure 6-76.

Surface Patterns, Texture, Ornamentation, and Details

In areas where there are pedestrians or slowly moving traffic, the use of patterns or texture on abutment walls can add interest. Any special patterns, textures, or materials should be complementary with nearby uses and buildings.

Historic railroad bridges often used the logo of the railroad company on the pier or painted on the girder. They demonstrated how visually helpful it was to break up these large areas of steel and concrete. Town logos, graphic designs inset in the concrete, small areas of bright color, and other devices can be used in the same way to add interest to these large and otherwise featureless structures.

Signing, Lighting, and Landscaping

Signs are less of a problem on a railroad bridge than they are on a highway bridge. The bridge is deep and bulky to begin with, and the sign is less likely to project above the upper edge of the structure.

A special situation exists where supports for electrical catenary must be carried on the bridge. If such a situation is unavoidable, the best approach is to widen the piers, and carry the catenary poles directly on the piers. Alternatively, the poles can be carried to the ground beside the piers. Supporting the catenary on the structure itself requires that both be studied together to develop a catenary support that looks like part of the structure and not a bolted-on afterthought.

If there will be pedestrian uses under and around the bridge, lighting may be a necessary feature for the underbridge area and its approaches.

Landscaping in the abutment area can be an effective way to reduce the apparent size of high abutments and should be seriously considered.

CASE STUDY

Humboldthaven Railroad Bridge, Berlin, Germany

After the reunification of East and West Germany, the German government undertook the reconstruction of the Lehrter Railroad Station. The project required the construction of 1,000 meters of elevated structure in an east–west direction with three different clearance conditions. The first required relatively small spans over cross streets and pedestrian/station uses with spans of about 20 meters. The second required spans of about 30 meters over a multi-platform subway station, which crosses under the elevated station in a north–south direction. The third was the crossing of the Humboldthaven barge harbor. The span required was 60 meters, with a height of about 10 meters above the water. The bridge had to accommodate six tracks in a curve on four independent structures, with the width of the gap in between varying to accommodate platforms, all covered by a 450-meter-long glass train shed (Figure 6-77). The immediate area around the bridge is near the new Reichstag and Chancellery, and is being redeveloped as housing and office space. Though the structure as a whole is a unique combination of different requirements, each part of it reflects conditions that occur frequently elsewhere, and all parts were dealt with in a consistent and innovative manner. Thus, it is a useful case study.

The design intention/vision for the structure was to create bridges that provided maximum flexibility for the various uses under the structure, opened up the views through the area as much as possible, and showed an organized, clean, and attractive underside in order to encourage the use of the station and the redevelopment of the area. At the same time, the overall structure had to appear as a single entity that harmonized these three different structural requirements (Figure 6-78.1).

The geometry was, of course, set by the requirements of the railroad. The basic decision was to use a solid concrete girder section about 1.7 meters deep throughout the bridge (Figure 6-78.2). This section is sufficient for the basic 20-meter spans. In the elevated parts, the girder is supported by tubular steel columns that branch transversly. Over the subway station, this girder is supported by tubular steel piers that branch both transversly and longitudinally (Figure 6-78.1). By branching the columns longitudinally, there, the effective girder spans are reduced to about 20 meters again. Over the Humboldthaven, this girder serves as the deck of a deck-stiffened tubular steel arch with tubular steel spandrel columns again about 20 meters apart (Figures 6-79.1 and 6-79.2). Throughout its length, the girder section provides both longitudinal bending resistance and sufficient torsional stiffness to accommodate the curved alignment. At the arch, the girder both stiffens the arch and provides sufficient torsional stiffness to allow for a substantial deck overhang, so the arch ribs can be placed closer together. The center section of the girder also serves as the keystone section of the arch. The columns arch ribs are formed from steel tubes with cast steel joint elements joined to the tubes

FIGURE 6-77 *The bridge at the Lehrter Railroad Station in Berlin uses a steel tube arch to cross the Humboldthaven barge basin. (See also color insert.)*

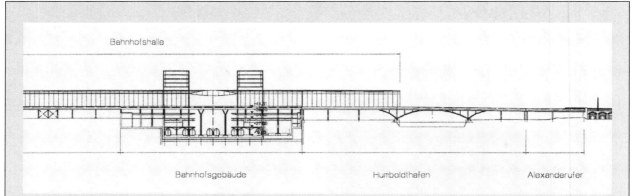

FIGURE 6-78.1 *The bridge must span a number of underbridge uses, including several streets and a rapid transit station, in addition to the barge basin, and serve as the base for the station's passenger platforms and roof.*

by welding. This use of cast–steel joint elements is a technique pioneered by the designer Jorg Schlaich, beginning with his structures for the 1972 Olympics; it is now starting to receive some attention in the United States.[2]

The bridge is not intended to attract attention to itself, so the pier shape is kept simple. The abutment shape is determined by the adjoining elements of which they are a part. The arches are painted medium gray, different enough in value to differentiate them from the concrete girder and to give them sufficient visual weight to match their role, but not so dark as to make the underside of the structure seem dark. There are no textures or ornamentation applied to the structure. The girders contain all conduit and drains. The only bracing for the arches is a simple and elegant cross brace in the plane of the pier columns. Elsewhere, sufficient stiffness is provided by girders, tubes, and cast steel joints so that no additional bracing is

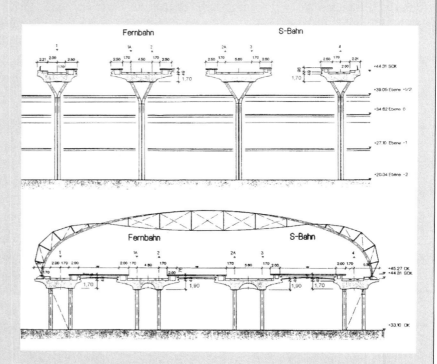

FIGURE 6-78.2 *The unifying feature of the bridge is a 1.7-meter-deep solid concrete girder.*

required. Similar cross braces provide lateral stiffness where required elsewhere in the structure (Figure 6–80).

The result is a bridge that clearly meets its design intention/vision. Open views are possible in all directions throughout the structure The wide overhang and relatively thin deck edges facilitate the penetration of light through the spaces between the bridges, and the light-colored surface of the uncoated concrete also keeps the underside light. In contrast to the undersides of most railroad viaducts, the underside here is an attractive place to be and should enhance use of the station and the redevelopment of the area.

The cost of the bridge was about 10 to 15 percent higher than a "standard" solution. The city planning authorities were adamant that the redeveloped area near the new station, with its high levels of pedestrian and automobile traffic, deserved an attractive bridge. Rather than lose an equivalent amount of money and a significant amount of time on restudies and controversy, the railroad authorities agreed to proceed with this design.

Schlaich Bergermann und Partners, of Stuttgart, designed the structure for the German National Railways.[3]

FIGURE 6-79.1 *The base section of the arch.*

FIGURE 6-79.2 *The cast-steel joint that forms the base of the arch prior to welding the tubes.*

FIGURE 6-80 *The only bracing of the arch is in the plane of the pier columns. Elsewhere, lateral stiffness is supplied by the tubes and their cast-steel joints.*

PEDESTRIAN BRIDGES

The following discussion is aimed at freestanding pedestrian bridges; it does not address pedestrian bridges that are built to connect buildings or parts of buildings. Because of their need to tie both physically and stylistically into the architecture of which they are a part, such structures require an approach tailored to each situation.

Developing a Design Intention/Vision

Physical Requirements

Span requirements are usually set by the clearance envelope of the undercrossing facility. Pedestrian approach requirements will be dominated by the regulations of the Americans with Disabilities Act (ADA), which sets ramp grades, widths, turning radii, and surface requirements. The nature of the crossing and its users may impose requirements for screens that prevent the dropping or throwing of objects on undercrossing vehicles.

Visual Environment

Pedestrian crossings can occur in any type of environment: residential, commercial, industrial, and undeveloped. If the area is developed, the facades of any nearby buildings will be important backgrounds. Often, pedestrian structures are associated with school playgrounds, parks, or open space greenway systems, where the background is formed by landscape features, trees, and grass.

The bridge will also be an important feature to people on the undercrossing roadway, whether in vehicles or not. There may also be areas where people congregate in the surrounding neighborhood that should be investigated as potentially important viewpoints.

Nearby and Associated Uses

Nearby land uses will determine the nature of typical users of the structure. Bridges near schools and sports facilities will be used by people in groups, who may

FIGURE 6-81 A pedestrian bridge that incorporates thinness, lightness, and a visual response to the concentration of forces.

well be in a boisterous mood. The need for security screens should be recognized. Structures associated with a park or art museum, in contrast, may have less need for such elements.

Symbolic Function

Pedestrian bridges are relatively lightweight structures, leading one to think that they would not have the presence to carry a major symbolic burden. However, there are many cases where a pedestrian bridge has been called on to do precisely that. For example, a nearby institution or sponsor, such as a corporation or museum, has often used a pedestrian bridge to make a statement about its activities and values. Engineers have used the opportunity to exhibit innovative structural concepts (Figure 6-82). Even in less-significant situations, the bridges are often very important to the life of a neighborhood, the residents of which value the symbol of safe connection provided by the structure. It is important to recognize all of these potential concerns and incorporate them in the design intent of the structure.

Boundaries

The structure cannot be considered independent of the pedestrian system that feeds into it. There are too many pedestrian bridges that are bypassed by their intended users because the ramps/stairs look too steep, the under-bridge area seems dark and threatening, or the approaches and/or structure seem too far off their intended direction of travel. Pedestrians are a contrary lot and they will seek the apparent shortest route between two points. The pedestrian bridge must look like it is that route. The bridge in Figure 2-13 took its U-shaped plan from the need to most directly reconnect the historic town's severed pedestrian main street.

The trick is to design the structure so that it either occupies, or seems to occupy, that shortest line between the two points. Grading, landscaping, and fencing can be used with pathways and the structure itself to create the impression that the structure

FIGURE 6-82 *Maillart's Toss River Bridge, in Winterthur, Switzerland, exemplifies structural art in a simple pedestrian bridge.*

is indeed the quickest and easiest way to get people where they are going. The view should at all times present an easily traversed path that is secure, interesting, inviting, and the most direct route to the intended destination. That means that the grading, landscaping, fencing, and approach pathways must be considered part of the bridge project. They must be designed and built at the same time.

Guidelines for Developing a Design Intention/Vision

Pedestrian bridges offer a unique opportunity to develop light and graceful structures. They look best when the superstructures are kept very slender and when the lines of the structure flow continuously over the supports and into the ramp or stair sections. Because stairs are inherently discontinuous features, real design ingenuity is required to integrate them smoothly into the balance of the structure. Ramps are much more easily accommodated. Railings and pedestrian screens can become heavy, over-done distractions, so it is best to keep them simple, transparent, and subordinate to the main lines of the structural members. When designing a pedestrian bridge:

- Create the visually most direct pedestrian route.

- Conceive the span and ramps/stairs together as one continuous structure,

- Strive for transparency, lightness, and continuity.

- Incorporate the symbolic aspirations of the community and/or sponsor.

Pedestrian bridges are the place to let the imagination run free.

Determinants of Appearance

Geometry

Because of the typically lightweight structure of the bridge, the geometry of the pathway will dominate the shape of the structure. Even with all of the requirements of the ADA and the need to present to users the appearance of a direct and efficient path, there is greater flexibility in the design of the pathway geometry than there is with roadways. The pathway geometry should be considered an integral part of the design of the whole structure and should be aligned to improve the appearance of the structure (Figure 6-83).

The structure will look best if the geometry flows with continuous gentle curves. If the space available imposes straight-line elements, then at least the connections between them should be curved. It is particularly important that vertical crest connections be long and continuous vertical curves. A kink in the vertical alignment will be annoying to both users and observers. With pedestrian bridges there is also the possibility of varying the width of the walkway. This can be done to accommodate the superstructure concept, to tie into branching pathways at the end of the structure, to provide an overlook opportunity on the structure, or just to enhance the visual interest of the bridge.

FIGURE 6-83 *Jörg Schlaich's Rosenstein I bridge, Stuttgart, Germany. The gentle rise of the deck of this suspension pedestrian bridge adds to its graceful appearance.*

Superstructure Type

Girders, trusses, arches, rigid frames, cable-stayed, and suspension bridges have all been used for pedestrian bridges in the span ranges that are the subjects of this book. Relative slenderness should be sought when choosing superstructure type. Continuity of structural form, material, and/or depth should be maintained.

Girders are the most simple and straightforward, but they can appear too heavy for their load, particularly if used in a deck configuration. The design of the parapet/railing/screen then becomes very important in establishing the impression created by the structure.

Trusses can be detailed to be light and graceful in appearance. Trusses using diagonals only (a Warren configuration with no verticals) and with all of the diagonals at the same angle are particularly effective. Removing the verticals and keeping all of the diagonals at the same angle reduces the visual confusion typical of trusses, and the diagonals are seen as one zig-zag line rather than a series of individual members. It is also important to simplify the wind bracing and sway bracing, again with the goal of reducing visual confusion. Trusses have the advantage that the pedestrian screen can be incorporated within the structural envelope, without requiring additional posts and fittings. Standardized, predesigned truss structures can be a very economical choice. Structures of this type are offered by a number of manufacturers, and, when detailed as just described, they can offer the same visual advantages as custom-designed structures.

If the site offers slopes or rock outcroppings that are natural places to found an arch or rigid frame, then these can be appropriate types of structures. At less-robust sites, tied arches or combinations of full and half-arches can be used. Another option is to use the abutment/ramp/stairs as opportunities to create massive elements that will contain an arch or rigid frame. If the arch can be kept flat enough, it is possible to place the pathway directly on the upper surface of the arch rib Figure 6-82 (page 275). Arches have an appealing gateway appearance that may be particularly appropriate in some situations.

Cable-stayed and suspension bridges offer the lightest possible appearance and the greatest flexibility in the horizontal geometry. They can be structures of exhilarating grace, but they require talented and sophisticated design and knowledgeable contractors. The pedestrian screen must be carefully designed to avoid losing the appearance of lightness.

Pier Placement

Pier placement will be determined by the needs of the undercrossing facility and the nature of the superstructure. These structures tend to be thin and light in weight, so they do not present the same sense of visual obstruction as other structures. This gives the designer greater flexibility in locating the piers, and in particular to place piers in logical relation to topographic features (Figure 6-84).[4]

FIGURE 6-84 *The Mount Vernon Tail Bridge in Washington, D.C., was aligned to avoid existing trees. The user walks or bicycles through the treetops.*

Abutment Placement

The abutments can be very important visual features for this type of bridge. The visual impression is sometimes dependent on the contrast between a light and airy structure and a massive abutment or approach ramp.

Because of the light weight of the superstructure, it is even more important than usual to keep the abutments far enough back from the edges of the pavement to maintain clear views through the structure for any undercrossing roadways. Otherwise, there will be a visual question: Why crowd the roadway edge with the abutments when a few more feet of apparently inexpensive superstructure would solve the problem?

Superstructure Shape

The goal is to present an appearance of lightness and thinness, therefore it is necessary to shape the superstructure in ways that achieve that end. It is important to seek dimensions and details that emphasize apparent thinness and horizontality and to keep structural depth constant or smoothly varied over the entire bridge. To keep the bridge light in appearance, concrete parapets should be avoided, unless the parapet is also the supporting girder. Use a full-height railing or pedestrian screen instead. Deck overhangs will be less on a pedestrian bridge, but can still have a significant influence in making the structure seem light and thin by casting a shadow on the girder below.

The railing/screen design will have a major impact on the impression created by the bridge. There are two reasonable approaches. The first is to keep the design as simple as possible, with the emphasis on horizontality. This approach is more compatible with arch and cable-supported structures. These superstructure types have strong shapes of their own with much intrinsic interest; Figure 6-83 (page 276) shows a good example. Elaborately detailed railing/screens can distract from the structural shape and diminish its appeal. On the other hand, girder and truss structures can seem bland, and elaborately detailed railings can add interest that would otherwise be missing. The approach then is to elaborate the screen design with details that become decorative motifs responding to the surrounding areas and/or the symbolic intentions of the structure. With truss structures, it is important for the details to be consistent with the basically triangular geometry of the truss.

Pier Shape

The piers also should present an appearance of lightness and thinness. This quality can be emphasized by chamfering the top of the pier and using high bearings to visually disconnect the pier from the superstructure.

The piers also offer an opportunity to create shapes that add interest to the structure. This is especially true and necessary for the towers of cable-supported bridges. The shapes must be consistent with the pier's load-carrying requirements, but that does not mean they have to be bland.

Abutment Shape

The abutment can be a very important visual foil for the superstructure. If the superstructure is a true arch, the abutment must provide the visual weight to convincingly contain the thrust of the arch. If the superstructure is a girder or truss, the abutment must support them in such a way that they seem to lightly touch down. For cable structures, the visual focus will be on the towers, and the abutments should be kept small and simple so as not to compete. The abutment is often required to serve as a stairway or ramp, or both, which should be gracefully integrated into the overall shape (Figure 6-85).

Color

Color is one economical way to add presence to a pedestrian bridge so that it can carry a symbolic design intention/vision. The color selected must have a specific symbolic and/or compatible relationship to the area and/or sponsor. One example would be using the colors of a nearby university for a structure used primarily by its students. Another would be selecting a color that is the complement to the dominant color of background woodlands (Figure 6-86). Simply by making the bridge an unusual visual event in an otherwise unnoticeable series of structures, color can convert the bridge into a neighborhood symbol. People will remember and remark on, for example, "the red bridge."

Surface Patterns, Texture, Ornamentation, and Details

Surface texture is a valuable way to add interest to and enhance the user's passage across the structure. Here the previously stated rules about large-scale detail are reversed. Pedestrian users will enjoy smaller features, particularly if they are amenable to touch. Likewise, ornamentation will be more likely to be appreciated at pedestrian speeds and distances. The elaboration of railing designs discussed previously is a type of ornamentation. Logos of towns or sponsoring institutions or abstract patterns are other possibilities. Lighting fixtures and posts can be ornamental in themselves, even when not lit. The fixture design should have a clear relationship to the design of piers and towers, so that the bridge looks as if it was designed all at once, and so that the fixtures do not appear to be afterthoughts.

Because the users are in such close contact with the pathway, the details of drainage may have more of an impact than on roadway structures. A roadway usually has a curb and gutter arrangement, which accepts drainage and debris and carries

FIGURE 6-85 *This abutment integrates both ramp and stairs.*

FIGURE 6-86 *The piers of the Rogue River Bridge in Grants Pass, Oregon, are placed to allow park uses to continue below. However, the railing color does not blend with the background, an example of the difficulty of finding a green that blends well with background foliage. A contrasting color might have been better (see color insert).*

them away. Even if some leftovers remain, they are seen as part of the street, not the sidewalk. On a pedestrian bridge, water and debris must be handled on the walkway surface. They often get trapped in pockets along the intersection between pathway and parapet, making the walk unattractive and perhaps impassable with ordinary shoes. Positive methods of handling drainage and debris must be provided. Leaving the edge open is the best way to get rid of both drainage and debris. If the edge must be closed, a small (4 to 6 inches) recessed gutter should be provided next to the parapet/railing, with frequent enough outlets to avoid overflows and trapped debris.

Signing, Lighting, and Landscaping

Signs can be successfully integrated into the design of truss pedestrian bridges—one of the few exceptions about signs on structures. The structure provides a depth that is of the same order of magnitude as the sign panel's vertical dimension. Also, the structure itself looks a bit like the familiar sign bridge.

Other types of pedestrian superstructures are not as hospitable to signs. Arch and cable-supported structures are particularly incompatible with signs. If a sign absolutely must be accommodated at a particular location, then that may be a presumptive reason to select a truss design there.

Lighting is a very important feature of pedestrian bridges, for both security and visual reasons. Unless the whole area is lit by high-mast fixtures, normal lighting on nearby roadways will be too uneven to be sufficient for users of a pedestrian bridge. Separate fixtures specifically designed to light the walkway both on and off the bridge should be provided. In designing this arrangement, it is necessary to avoid glare for both walkway users and drivers on the undercrossing roadway. Sharp cut-off fixtures should be used. The type of light source used (mercury vapor, high-pressure sodium, etc.) should be selected to complement the materials and colors used in the structure. Lighting built into the railing is particularly effective.

Landscaping can be a valuable adjunct to pedestrian structures, as it can help to direct users toward the desired route, provide additional interest and pleasure for users, and complement the driver's views of the structure.

FIGURE 6-87 *Railings can be made of cables and mesh, and retain a light appearance.*

CASE STUDY

Reedy River Pedestrian Bridge, Greenville, South Carolina

Greenville's downtown is split by a wooded valley park that contains the falls of the Reedy River. The falls were bridged by a street crossing that effectively obscured and shadowed them, as well as interrupted the park. The city wanted to redevelop the park as a botanical garden and to strengthen the pedestrian connections between the park and the downtown along the Reedy River. The decision was made to remove the street bridge from the park and replace it with a pedestrian bridge located just downstream from the falls to maintain the pedestrian connection between the two sides of the downtown. The pedestrian bridge was to be 12 feet wide, with a minimal number of supports in the park.

The visual environment is dominated by the deciduous trees of the park, though the nearby buildings of the downtown are also visible, most of which are two- to five-story high commercial and office buildings in various shades of brick. There is significant pedestrian use of the park, which the city hopes to increase. The bridge will stand at a prominent location within the city and be visible from several nearby streets. The boundaries of the bridge are set by the wooded slopes of the valley.

The design intention/vision for the pedestrian bridge included two goals. On the one hand, it must have enough height and presence to serve as a landmark for the downtown and the new botanical garden. On the other hand, when seen from the park, it must not overshadow the park or interfere with views of the woodlands or the surrounding community.

The deck of the Reedy River pedestrian bridge will be placed on a curve to tie into the streets/plazas on either side and to provide the best viewing position for the falls (Figure 6-88). The concave side of the deck curve faces the falls, creating an aerial amphitheater from which to view the cascading water. The curve also allows a one-sided suspension system, which places all of the suspenders on the side away from the falls. The bridge is a curved suspension bridge with a total length of approximately 380 feet and a clear span of 200 feet. The single cable is supported by two towers placed in the trees well back from the river's edge The twin steel tubular towers are inclined away from the bridge and stabilized by backstay cables.

The superstructure of the bridge consists of a thin concrete deck supported by elegantly shaped triangular steel frames that provide the anchoring points for the suspender cables. The structural system is completed by a curved ring cable below the deck, which is tied to the abutments and attached to the steel frames. This cable provides the torsion necessary to balance the main suspension system as well as control dynamic loads. These elements create a three-dimensional truss structure that will be stable,

FIGURE 6-88 *The Reedy River Bridge in Greenville, South Carolina, will connect the two halves of the downtown across a river valley park and provide a platform for viewing the falls of the river.*

light in weight and very transparent (Figure 6-89). The pedestrian railings are shaped to complement the steel and are curved in shape. They will be fabricated of stainless steel for durability and visual appeal. A transparent stainless-steel mesh will be hung from the railing and will follow the curvature of the bridge to provide safety for users and visually enhance the bridge.

The abutments are the minimum size needed to adapt the end of the deck to

FIGURE 6-89 *The cable and deck elements of the bridge will combine to form a three-dimensional structure that is stable, light in weight, and transparent.*

the slope and provide an anchorage for the cables. The towers and cables will be painted white. The railings will be unpainted stainless steel. The top rail will be shaped to allow for the integration of light-emitting diode (LED) light strips, which will illuminate the deck surface at night. The main structural components of the bridge will also be illuminated at night; in particular, the two towers will be accented to create a dramatic appearance in the downtown (Figure 6-90).

When seen from below, thanks to its thin deck and spidery suspension system, the new bridge will appear to float through the treetops. At the same time, the twin towers and suspension cable will be visible from vantage points around the city, calling attention and drawing visitors to the public botanical garden, falls, and river. Thus the bridge will meet all aspects of the city's design intention/vision.

Schlaich Bergermann und Partners, of Stuttgart, Germany, developed the design in a joint venture with Rosales Gottemoeller & Associates. Andrea Kratz was designer, and Miguel Rosales was lead aesthetic advisor. Construction started in September 2003 and is expected to finish in the summer of 2004.

FIGURE 6-90 *The Reedy River Bridge will be lighted to be a landmark for Greenville at night as well as during the day (see color image).*

1   This case study is based on an unpublished paper entitled "Bridge Aesthetics for the I-25/I-40 (Big I) System Interchange Reconstruction," presented by Alex Whitney of the URS Corporation at the 2004 meeting of the Transportation Research Board, Washington, D.C.

2   Schober, Hans, "Steel Castings in Architecture and Engineering," *Modern Steel Construction* (April, 2003). No page info. on reprint

3   For more on the work of Jorg Schlaich, see Holgate, Alan. 1997, *The Art of Structural Engineering: The Work of Jorg Schlaich and His Team* (Stuttgart/London: Axel Mendes).

4   The author was aesthetic advisor to HDR Engineering, Inc. for the design of the Mount Vernon Trail Bridges at Reagan National Airport, Washington, D.C.

chapter *seven*

MAKING IT HAPPEN: ORGANIZATIONAL CONSIDERATIONS

"When the design philosophy from top management to the actual design team is one that encourages beautiful structural design, the end result can be an aesthetically pleasing structure that is also economical."

—JAMES E. ROBERTS, FROM *ESTHETICS IN CONCRETE BRIDGE DESIGN.*[1]

The dreary appearance of many everyday bridges is not solely due to a lack of good intentions among designers. Most engineers do believe that concern for appearance should be an integral part of their work, but improvement may be beyond their powers as individual engineers. Engineers respond to the priorities of their clients—governmental highway departments, toll authorities, public works departments, and transit agencies—which may or may not have made appearance a priority. Likewise, their elected officials and their public may not have given clear direction that appearance is important. Finally, their engineering education may not have given them the wherewithal to respond. This chapter addresses how to resolve those issues and the particular contribution that engineering design competitions can make.

THE RESPONSIBILITIES OF BRIDGE-BUILDING ORGANIZATIONS

Governmental highway departments, toll authorities, public works departments, and transit agencies are the key decision makers when it comes to the quality of America's bridges. They set the standards, select the designers, and pay for the results. Each of these agencies has an aesthetic policy, whether it realizes it or not. Just as every bridge produces an aesthetic impact, each agency's bridges, taken together, enunciate that agency's aesthetic policy. It may be a policy of apathy or ignorance, but it has its effect nevertheless (Figure 7-1). The relevant issue about an aesthetic policy is not whether it exists, but the quality of the bridges it produces.

A number of bridge-building agencies have made appearance an integral consideration in their bridge building. It is an integral part of their planning, preliminary design, final design, construction, and maintenance. Concern for aesthetics becomes

FIGURE 7-1 *The result of a policy of aesthetic apathy or ignorance.* 　　**FIGURE 7-2** *The result of a policy of knowledge and concern.*

a routine part of job descriptions, requests for proposals, consulting contracts, standard details, and employee and consultant evaluation. These agencies face all of the same limitations of time, money, and political interest as any other, yet they are able to build bridges that are attractive assets in their communities. Many of the better examples in this book are not special structures built in special circumstances; they are typical products of these agencies, built to the same high standards as all of that agency's bridges (Figure 7-2).

Perhaps an individual engineer working as an employee or consultant cannot expect to change an agency's aesthetic policy, though he or she should certainly take advantage of opportunities to influence it. However, the engineers/managers running an agency do have the power to improve an agency's aesthetic policy, and should do so. Unfortunately, engineer/managers often cite one or more of the following reasons why they don't. Let's examine each one.

- *The bridges will cost more.* Not necessarily true, as we saw in earlier chapters.

- *The bridge design will cost more.* Again, not necessarily true, once a training and transition period takes place.

- *With our production schedule and/or staff levels, we can't afford the time to train our staff and/or revise our procedures.* There is an unavoidable cost in time and money to train and install new procedures and standards. However, if this is an insurmountable obstacle, one must ask whether the agency intends to *ever* make *any* improvements for *any* reason. Most agencies have some budget allowance for staff development and procedural improvement. If these resources were focused on aesthetics for a short period, much could be accomplished with little disruption to established budgets and schedules.

- *The public/governor/legislature/mayor/council do not care about better-looking bridges.* They probably don't know that they can *have* better-looking bridges. They have been numbed by mediocrity. Any active bridge-building agency will have had experience with situations in which the appearance of a bridge became controversial. These instances prove that people care. Good appearance is part of the value they expect to get for their tax money. This is also something that the engineer manager can do something about, as discussed below.

There are too many easy-to-find examples in which excellent aesthetic policies have been established within all of the usual constraints of budget and politics for any engineer/manager to say it cannot be done. Again, one can cite organizations as different as the California Department of Transportation, the Tennessee Department of Transportation, the Ontario Ministry of Transport, and the Washington, DC, Metro, which have done exactly that. Engineer/managers in public agencies have a professional responsibility to the public to bring the best practice of the profession into their agencies. For the same reason that they recommend the adoption of new materials or improved construction techniques when they believe them to be cost-effective, engineer/managers should recommend the best aesthetic policies for their agencies.

Americans want full value for their tax money. They want not *just* bridges, but *beautiful* bridges. It is up to engineer/managers to find ways to satisfy that desire. Here are some ways to do so.

Educate the Public and Elected Officials

Americans have not been clear concerning their desire to have visually attractive public works. Calls for better-looking structures often generate ill-informed responses alleging better-looking structures cost more, leaving politicians and public works officials indifferent or actually hostile to concerns about aesthetics.

The basic problem is a general perception that improved appearance will automatically cost more. The experience from states such as California, Tennessee, and Washington, and the Province of Ontario proves that this belief is mistaken. Excellent appearance can be achieved with cost-effective design. These agencies do it routinely (Figure 7-3). Staff, consulting engineers, user groups, contractor groups, elected officials, and the public need to be educated to this fact. Engineer/managers should be spreading the word to elected officials that quality appearance does not have to cost more.

Groups interested in scenic quality and urban design are potential allies in this effort. These groups are often squelched on the basis that "mere aesthetics" should

FIGURE 7-3 *Routine excellence in North Carolina.*

not rank with matters such as education and public safety in the competition for tax dollars. If good appearance need not cost more, then it is no longer a budget issue. That puts these people in a stronger position to request a higher level of performance from public works agencies.

Improve the Agency's Aesthetic Policy

There are four steps to improving an agency's aesthetic policies:

1. Establish aesthetic quality as an explicit goal of the agency.
2. Upgrade the skill of engineers and consultants.
3. Remove bureaucratic barriers.
4. Review standard details and procedures.

Establish Improved Aesthetic Quality as a Goal

An agency usually gets what it asks for. In order to get better-looking bridges, an agency must ask for them. The first step is to announce that improved aesthetic quality is an explicit agency goal. The second step is to make improved aesthetic quality an explicit criterion for every bridge under design, with the clear understanding that the improvement is to be made within the agency's current cost structure.

The announcements must be communicated in an effective way, then reinforced through seminars, conferences, and newsletters. Aesthetic quality has to become an integral part of employee recognition. Requests for consultant proposals, consulting contracts, evaluations, and design procedures must also explicitly incorporate aesthetic concerns.

Improve Skills

Training programs need to be established to improve the aesthetic abilities of staff design engineers and consultants working for the agency. This step is key, for it is from the efforts of engineers working on specific bridges that guidelines, details, and, eventually, policies will evolve. Seminars, in-house mini-competitions, and case studies of recent agency bridges can all be valuable teaching tools.

Remove Bureaucratic Barriers

The typical methods of organization within an engineering agency impose barriers to aesthetic improvement. For example, the design of a bridge is usually broken down sequentially among different groups of engineers. One group may do the preliminary design. A second group may do the final design. A third group will review it. All will make some changes during this process. A fourth group of engineers may develop some of the details, such as parapet designs, which may be applied years later to bridges that were not even conceived at the time the details were developed. As a result, it is difficult for a unified concept to emerge, hence often the result is a visual hodgepodge. The solution is to place more of the total design of each structure in the control of one person or group, including the ability to modify standard details if the situation calls for it.

Problems often arise even before the design of the bridge itself begins. The position and alignment geometry of a bridge is usually determined beforehand, by a highway engineering division. Often, this geometry is treated as fixed for the purpose of bridge design. However, there are often small changes in the vertical and horizontal alignment that would make major improvements in the appearance of the bridge at no significant detriment to the highway design. Because of the split between highway and bridge design functions, bridge engineers may not be able to implement these changes. The placement of signs on bridge structures is often handled the same way, without an opportunity to find alternate locations. Bridge design procedures should be changed to give bridge designers an opportunity to offer feedback to the highway designers and others and, subsequently, obtain needed changes.

Review Standard Details

Perhaps the single biggest impediment to improved aesthetic policy is the inertia created by inappropriate standard details. Because small- and medium-span bridges are so numerous, many agencies view their design as a process of assembling standard details. Standard pier shapes, parapet profiles, and standard abutments essentially establish the appearance of an average highway overcrossing no matter what else the designer might do. There is some obvious economy in this approach. However, it does not relieve anybody of responsibility for the appearance of the resulting bridge. There are many agencies that use standard details to achieve excellent performance, reasonable cost, and outstanding appearance, *all at the same time.* They do it by developing attractive details to start with, then allowing their designers the flexibility to apply them at the appropriate place. California has approached this problem by developing a series of standard attractive pier shapes for use with their typical concrete box girder bridge. The series provides for variety—rounded versus faceted designs, for example—but it is limited to a manageable number of variations. Knowing that designs will almost always come from one of these standard shapes, contractors and form rental companies feel safe in investing in the specialized forms required. Since these forms can be written off against many jobs,

FIGURE 7-4 *California bridges such as this typical example benefit from the application of attractive standard pier shapes and other attractive standard details.*

their cost ceases to be a factor in deciding which shape to use. The result is bridges that are both attractive and economical.

The basic truth is this: A standard detail can save money regardless whether it is attractive or ugly. Why not make it attractive? Why grind out mediocrity when, for the same price, you can grind out beauty?

An appearance criterion must first be applied to the standard details. Do they produce attractive bridges? Then it must be applied to the use of the details for each specific bridge. Are the details appropriately applied to this particular bridge, or should they be modified or disregarded?

An agency should take a comprehensive review of the effect on appearance of all of its standard details. Then each can be revised to contribute to the overall aesthetic concept that the agency is seeking to develop and to give designers more flexibility. While the functional aspects of standard details must be respected, their appearance should be reconsidered for each bridge to make sure they fit that bridge's design intent.

Ask Educators to Teach Bridge Aesthetics

If aesthetic quality is a legitimate criterion for bridge design, than teaching how to achieve it is a legitimate part of the engineering curriculum. Agency engineer/managers, as major employers of engineers, are in a strong position to influence curriculum writers. This is particularly true for public universities, which answer to the same governor and legislature as the state agency. Agency engineer/managers can also help by supporting university requests for funds and grants, by offering presentations based on their experiences with particular projects, by mentoring engineering students, by supplying data for case studies and design assignments, and by participating in student design juries and presentations.

THE RESPONSIBILITIES OF PROFESSIONAL ORGANIZATIONS AND CONSULTANTS

The best engineers, going back to the founder of the profession, Thomas Telford, have made aesthetic quality an integral part of their work. They have shown that aesthetic quality is indeed a legitimate engineering criterion, as essential as performance and economy, and that it is not to be delegated to others. It is time for the professional societies and consulting organizations to embrace the standards of their forbears and support the pursuit of aesthetic quality in all engineering activities.

Educate the Public and Elected Officials

Engineers and consultants outside of government have more flexibility in proposing new ides than engineering/managers of public agencies. The very nature of their profession gives them credibility with the public and elected officials. They are in a position to be opinion leaders in debates over agency policies and solutions for specific bridge sites. Many engineers travel and read widely, and know what is being done in other places. They may have even more knowledge than agency engineer/managers about the successes that have been achieved elsewhere. Engineers should put their knowledge and credibility to use, and seek to influence public decisions in favor of more attractive bridges.

Support Building Aesthetic Skills

Many of the activities outlined in Chapter Two for improving aesthetic skills would benefit from the support of engineering societies and consultant firms. Activities such as seminars and conferences on bridge aesthetics can be promoted. Many state registration authorities are requiring continuing education. Courses in bridge aesthetics are a legitimate response to these requirements.

Ask Educators to Teach Aesthetics

Likewise, professional organizations and consulting firms, also major employers of engineers, are in a strong position to influence curriculum writers. Individual professionals can also help by offering presentations based on their experiences with particular projects, by mentoring students, and by participating in case studies and juries for design assignments.

THE RESPONSIBILITIES OF ENGINEERING EDUCATORS

School is the natural place to lay the foundation for aesthetic skill. However, in the struggle for a place in the curriculum, aesthetics may not seem as necessary as strength of materials and partial differential equations. Nevertheless, the public holds the engineering profession responsible for all of its decisions, including those that affect appearance. When measured by the number of people affected, aesthetic decisions can be very serious indeed. How will this responsibility be discharged if no basis for this decision making is developed in the schools?

Some schools are beginning to respond to public criticism of the narrow base of engineering education by enlarging the place of the humanities in the curriculum and by providing "Introduction to Engineering" courses, which bring forth all of the varied aspects of the engineer's role. Consideration of aesthetics has a natural and indispensable place in both of these efforts.

Other schools have had success by introducing aesthetic considerations in senior-level design courses. A typical bridge is assigned as a design project. Sometimes the local highway agency supplies data on a project current in its program. This lends realism to the project, allowing students to view the site in the field and to take advantage of traffic, geological, and drainage data that the agency has assembled. The students are asked to develop a design for the entire bridge. Consulting engineers and contractors participate as advisors. The students often work in groups, in a mini-competition format, with the final designs reviewed by a jury of faculty, agency staff, and consulting engineers. Aesthetics is a consideration, along with structural effectiveness, economy, and constructability. During the project, students are exposed to, and learn about, the entire range of issues present in actual bridge projects, including aesthetic quality. The excitement generated by the intergroup competition makes the whole experience additionally effective.

Such introductory courses and design exercises are a beginning, but they don't do enough to get at the root of the problem, which is the almost-exclusive focus on analysis and the almost-complete avoidance of synthesis in the way engineering is taught. Engineering problems from the very first assignment in statics are presented as a series of loads, along with a predetermined system to resist the loads, usually drawn as a series of lines and symbols.

The student's job is to determine the moments and forces in the system. In more advanced "design" courses, the student is asked to size the members of the predetermined system. Rarely is the student asked to create (synthesize) the system itself, let alone given any advice on how to do that, or how to evaluate the results. Synthesis is the heart of design, but it rarely appears in the American engineering curriculum.

David Billington believes a more basic reorientation of the engineering education is necessary. In his book *The Art of Structural Design, a Swiss Legacy*,[2] he draws a relationship between the success of master engineers such as Robert Maillart, Othmar Amman, and Christian Menn, and the teaching tradition at the Swiss Federal Institute of Technology (ETH) in Zurich, from which they all benefited (Figure 7-5). In the book he describes lectures that are as much about synthesis as analysis, and which include specific discussion of the visual results of different solutions as well as their performance and economy. For example, a lecture on arch structures would discuss the performance and constructability of various options and their desirable visual proportions, as well as how they could be analyzed mathematically. The analysis would be approached with full mathematical rigor, but the discussion would also include simplified analytical methods to make it easier for engineers to investigate multiple variations. In describing the benefits of this tradition, Billington points to the one American work by Menn, the Zakim Bridge (cover), saying:

> *His American bridge signals a new challenge to educators in the United States. Works like . . . the Leonard P. Zakim Bridge all stand for the potentials for an improved environment through structural art. The age of technology makes possible the art of the engineer who can fuse the efficiency of form, economy of constraints, and elegance of construction into symbols of conservation, accountability, and aesthetics.*

FIGURE 7-5 *The Gantor Bridge on the road to the Simplon Pass in Switzerland, by Christian Menn, a product of Swiss engineering education.*

THE CONTRIBUTION OF ENGINEERING DESIGN COMPETITIONS

Occasionally, a situation arises in which the problems of structural conditions, public exposure, and/or environmental impact seem to exceed the capabilities of the usual bridge designs. In Switzerland, Germany, and France, such situations are often resolved by holding engineering design competitions. In these competitions, appearance is made an explicit criterion, along with performance and cost. The competition attracts outstanding engineers who develop designs that are often much better than would otherwise be seen. Because of that fact, and because the community can get involved in picking from alternatives, the competition often successfully resolves stubborn controversies.

The designs also have influence beyond the competition. The concepts influence subsequent projects at other sites, and thus influence the development of the profession. With appearance as an explicit criterion, competitions have proved instrumental in sensitizing engineers to aesthetics and to raising the general level of bridge appearance in these countries.

In the United States, unfortunately, competitions are rare. Most U.S. agencies resolve controversial bridge projects by a costly, time-consuming process. They develop successive alternative designs until one is found that the public will support. In contrast, Maryland has successfully resolved two controversial bridge replacements, for the Severn River adjoining the U.S. Naval Academy in Annapolis and for the Woodrow Wilson Bridge over the Potomac River at Washington, DC, by holding engineering design competitions.

The rules for the two competitions were similar and were based on the European model, though with a larger role for public groups and elected officials than is customary in Europe. The competition's prize was significant: The winning design firm would be contracted to perform the final design for the bridge. In addition, a certain amount of prize money would be awarded. Finally, each finalist would be paid a fee to prepare its entry. In view of the state's desire to select the bridge designs by competition, the Federal Highway Administration agreed to accept the winning concept and not require alternate designs.

The competitions began with advertisements in the engineering press and local newspapers. The ads invited qualified engineering firms to submit letters of interest outlining their qualifications and previous work of note. Numerous letters of interest were received. Based on these credentials, six firms were chosen as finalists for the Severn River project and four were chosen for the Woodrow Wilson Bridge. The finalists were each asked to prepare a preliminary design for the bridge using design criteria defined by a written "Program and Rules" document. The design had to be prepared in sufficient detail to enable definitive decisions about the sizes of the major structural components of the bridge and preparation of a realistic cost estimate. All of the drawings and other submission materials were identified only by code letters so that a specific proposal could not be linked with a particular entrant.

Juries were formed, made up of prominent engineers, architects and artists, representatives of local governments, interested state and federal agencies, and community groups. The juries were advised by review panels consisting of experts in bridge engineering and construction and, in the case of the Woodrow Wilson Bridge, urban

FIGURE 7-6 *The Naval Academy Bridge over the Severn River, Annapolis, Maryland. The result of a competition that resolved a major local controversy (see color insert).*

design. The juries selected the winners (Figure 1-22, page 19, and Figure 7-6), ranked the other proposals, and awarded the prize money. Cost was an explicit criterion in the selections. Based on the juries' recommendation, the State Highway Administration awarded the contract for the final design to the winning firms, which immediately began work.

The initial public reactions to the selections were almost all positive. Because of the broad-based membership of the juries, all of the relevant elected officials were able to support the selection. When critical statements were received at a later date from a few narrowly focused groups, the elected officials' consensus remained firm. When the critics later brought suit, the courts ruled that the processes met all necessary legal requirements.

The competitions successfully resolved the longstanding controversies at each site. Even more important, they identified two excellent bridge designs. The Severn River Bridge was completed in 1994 to general acclaim, even by many who had opposed the original proposal. The Woodrow Wilson Bridge competition is described in more detail in the following case study. The Woodrow Wilson Bridge is now under construction. It remains to be seen whether either bridge will influence the design of future bridges.

CASE STUDY

Woodrow Wilson Bridge Design Competition, Washington, D.C.[4]

The Woodrow Wilson Memorial Bridge is the only Potomac River crossing in the southern half of the Washington, D.C., metropolitan area. It carries the Capital Beltway, I-495, and the main north/south interstate route on the East Coast, I-95, across the river. Built originally with six lanes to carry 75,000 vehicles per day, it now carries 175,000 vehicles per day and is expected to carry 300,000 vehicles per day by 2020. The bridge includes a movable span to accommodate ocean-going shipping to Alexandria and the District of Columbia. Built in the early days of the development of the Interstate System with an emphasis on economy and little concern for appearance, the existing bridge has never been seen as a civic asset by the communities on either side (Figure 7-7).

On the Virginia side, the bridge approaches cross the Old Town Historic District of Alexandria, a river port town founded in the eighteenth century. This district includes many buildings that date from the eighteenth and nineteenth centuries, some of which are within sight of the bridge. The approaches also cross Jones Point Park, site of an eighteenth-century Potomac River lighthouse. On the Maryland side, the approaches cross Queen Anne's Park and adjoin National Harbor, designated a major development site for Prince

FIGURE 7-7 *The existing Woodrow Wilson Bridge as seen from Jones Point.*

Georges County. The bridge is visible from the White House and many other locations in Washington, and is considered part of the city's monumental core. The replacement of the bridge raised many concerns on the part of both community groups and review agencies about the visual and urban design effects of the replacement bridge on these historical, architectural, and community resources.

In 1987, the Federal Highway Administration (FHWA), Maryland, Virginia, and the District of Columbia began a series of studies to determine "the best alternative to enhance mobility in the corridor while addressing community and environmental concerns." As the studies progressed, the City of Alexandria, Prince Georges County (Maryland), Fairfax County (Virginia), and the National Park Service were added to the committee overseeing the studies. Urban design studies and extensive community participation efforts were included in the scope of the studies to ensure that community, architectural, and urban design issues were considered, along with traffic, structural, and environmental issues.

This work culminated with the selection of a comprehensive alternative that included two parallel, six-lane bridges with movable spans at the navigation channel, along with significant highway improvements to the interchanges at both ends of the bridge. The replacement bridges will occupy a new, wider right-of-way just south of the existing bridge, which will allow the existing bridge to stay in service while the new bridges are being built. The new bridges will be separated by 15 feet, to allow light to penetrate to the under-bridge areas. The selected alternative also includes major urban design, housing, park and environmental components, created at least in part to mitigate and reduce the impacts of the new, wider bridges. The study articulated the following design intention/vision for the bridge:[3]

- Spans as long as possible, to open views up and down the river and to improve the usefulness and attractiveness of Jones Point Park.

- Arches or an archlike appearance to maintain the tradition of Washington's Potomac River bridges (Figure 7-8).

- A requirement that the bridge respect the historical tradition of Alexandria.

- Enhancement of the areas under the bridge for uses associated with Jones Point Park.

- Maintenance of vistas down certain streets in Old Town Alexandria.

- Continued memorialization of Woodrow Wilson.

- A pedestrian bikeway facility on the bridge to connect the Mt. Vernon Trail on the Virginia side with the Potomac Heritage Trail on the Maryland side.

FIGURE 7-8 *The Arlington Memorial Bridge is what comes to mind for most people when they think about bridges over the Potomac River in Washington, D.C.*

Following the completion of this study the FHWA, Maryland, and Virginia agreed to select the design for the bridge via a design competition. The competition used an approach similar to Maryland's successful 1988 design competition for the prize-winning U.S. Naval Academy Bridge in Annapolis. In January 1998, a general invitation was issued for interested groups to submit their credentials.

Seven teams submitted credentials; in April 1998, four were selected as entrants. Detailed rules were given to the entrants that set forth the physical requirements on the structures and made it clear that the proposals had to respond to the urban design criteria listed, as well as to structural and cost criteria. The entrants were permitted to submit two proposals each, at least one of which had to display an archlike appearance. The proposals also had to include sufficient detail, including calculations, to confirm the sizing of major members and permit preparation of a cost estimate. The entries were to be submitted anonymously, so that a specific proposal could not be linked with a particular entrant. The entrants were given 10 weeks to prepare their designs, and received $100,000 each to support their costs.

Seven designs in total were submitted by the four entrants.[4] The designs covered a wide variety of ideas, ranging from sophisticated applications of post-tensioned concrete to fairly standard applications of welded steel plate girders. (See Figures 7-9.1, 7-9.2 and 7-9.3 for views of three of the submissions.) The designs were reviewed by a citizens' advisory committee and three teams of independent specialists, covering the areas of structural design, cost and constructability, and historic and cultural resources. These reviews were passed on to a jury of 15, which included elected and appointed officials from the area and independent professionals in the fields of engineering and architecture. The jury was chaired by former Maryland Governor Harry R. Hughes. In November 1998, the jury met, reviewed the proposals, and made its selection. Entrant B, the design created by the Parsons Transportation Group (PTG), was declared the winner (Figure 7-9.2). The author's firm, Rosales Gottemoeller & Associates, was the urban design and aesthetic advisor to the PTG competition team.

One of the key components of Entrant B was a proposal to raise the Maryland end of the bridge by 10 feet to allow for the continuation of longer spans all the way to the Maryland shore. In general, multispan bridges look better if the span-to-height ratio remains constant from span to span. This has the effect of reducing the spans as the distance above the water decreases. The profile suggested with the competition documents dropped quite close to the water level on the Maryland side. This would have significantly shortened the spans on the Maryland side, introducing additional piers and blocking views through the structure, as can be seen in Entrant A's submission (Figure 7-9.1, right side). Raising the profile kept the proportions consistent but significantly reduced the number of piers and produced a more open appearance all of the way to the Maryland shore.

One of the most basic visual issues of the project was whether to visually differentiate the movable span, as in the existing bridge, or to make the movable span as similar as possible to the other spans, so that the bridge would visually flow from shore to shore. One of the more prominent of the traditional Potomac River bridges, the Arlington Memorial Bridge (Figure 7-8, page 296), has a movable span that is very similar to its approach spans. With this as a model, the movable span of Entrant B was made as similar as possible to the approach spans (see Figure 7-11, page 298).

The requirement that the bridge use arches or have an archlike appearance was one of the major criteria of the competition. Unfortunately, the foundation conditions at the site do not lend themselves to arch construction. The Potomac River bed is formed of alluvial soils that descend as much as several hundred feet below water level and that have very poor bearing capacities. The usual method of building a string of arches in these conditions is to carry the horizontal reactions from pier to pier until they reach the abutments. However, in this case, the movable span interrupts the string at the navigation channel, where the foundation conditions are the worst. Building the structure as a string of true arches would have required the construction of sizable and expensive foundations at the movable span. Indeed, there was concern about whether the movable span piers would be sufficiently stable against lateral movement, even with massive foundations.

Entrant B resolved these interlocking and seemingly contradictory requirements by building the bridge on a series of V-shaped piers supporting continuous haunched steel girders (Figure 7-10). With this system, the lateral forces on the piers are reduced to primarily the wind and seismic loads that would be present in any case. There are no arch forces. The arms of the V are curved, and visually interact with the soffits of the haunched girders to form a continuously curved line that emulates a series of arches. The V-pier requires a tension tie at the top, between the arms of the V; but the tie is placed between adjacent girders, so that it is visible only from directly below.

Clear views through the bridge were an important part of the design intention, but the great width of the bridge, 250 feet, makes it difficult to open vistas through the bridge; the wide substructure interferes with diagonal view lines. The desire to open up views through the structure led to a decision to seek long spans, with a goal of achieving a maximum span of about 400 feet. Even that span is only about 1.6 times

FIGURE 7-9.1 *Woodrow Wilson Bridge design competition, Entrant A.*

FIGURE 7-9.2 *Woodrow Wilson Bridge design competition, Entrant B.*

FIGURE 7-9.3 *Woodrow Wilson Bridge design competition, Entrant X.*

FIGURE 7-10 *Entrant B's design for the Woodrow Wilson Bridge, in Jones Point Park.*

the width. Because of the unique pier design for Entrant B, the strict proportionality of the spans could be relaxed, opening up the views still further. The span-to-height ratio of each V-pier is kept the same throughout, giving the structure a sense of consistent proportionality. However, the spans between the V-piers decrease at a lesser rate than the height, so that the total span from the base of one pier to the base of the next decreases at a slower rate than the height declines. The result is that the span-to-height ratio of the spans nearest the Maryland shore is greater than for those near the channel. With longer spans near the Maryland shore, the diagonal views through the bridge are maintained. The visual effect is to reduce the apparent complexity of the structure and open up view corridors through the bridge from every viewpoint (Figure 7-12). The proposed concept has 11 pier lines across the river, versus the 33 in the existing bridge. The longest river span is 402 feet; the shortest, 300 feet. Comparable spans continue through Jones Point Park, allowing maximum use of the park area under the bridge (Figure 7-10). In a further effort to open up oblique view lines, the competition team decided to maximize the deck overhangs and thus reduce the width of the substructure. The V-piers also flare outward in the transverse direction beginning from a narrow base. The narrow base means they are less likely to block diagonal views.

FIGURE 7-11 *Entrant B's design for the Woodrow Wilson Bridge, movable span. (See color insert.)*

The V-shaped pier is adapted at the movable spans to house a double-leaf bascule bridge. The result is that the movable span is very similar in appearance to the approach spans (Figure 7-12). Because the channel is very close to the Virginia shore, the operator's house could be located in a tower on the Virginia shore, where it would be hidden behind the tree line from most viewpoints but still have excellent sight lines to the channel. This solution would also provide public elevator access from Jones Point Park to the bridge sidewalk.

The jury's comments on the concept focused on its open appearance created by the large spans, and the way the movable spans blend into the balance of the structures. The graceful curves of the V-piers were also recognized. One jury member said, "The piers appear to spring from the water, like Neptune's hand holding the bridge." Also recognized were the relatively simple foundations, ease of construction, and

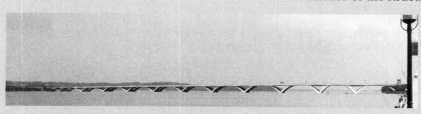

FIGURE 7-12 *Entrant B's design for the Woodrow Wilson Bridge, from the Alexandria Waterfront.*

relative economy the V-piers offer. One jury member said it was the entry with the "best opportunity to be completed on time and within budget."

The design contract was awarded to the winning entrant, the Parson's Transportation Group, and is now complete. The bridge is under construction. It is expected that the eastbound bridge will be complete and open to traffic in 2006; anticipated completion of the entire bridge is in 2008.

1   Roberts, James E. (Steward C. Watson and M.K. Hurd, eds.), 1990. *Esthetics in Concrete Bridge Design.* Detroit, MI: American Concrete Institute.

2   Billington, David P. 2003. *The Art of Structural Design: A Swiss Legacy,* Princeton University Art Museum. Princeton, NJ.

3   U.S. Department of Transportation, Federal Highway Administration, and Virginia Department of Transportation Maryland Department of Transportation, State Highway Administration and District of Columbia Department of Public Works, 1997. *Woodrow Wilson Bridge Improvement Study: Final Environmental Impact Statement,* Washington, DC.

4   Maryland Department of Transportation, State Highway Administration, 1999. *Woodrow Wilson Bridge Design Competition,* Baltimore, MD.

conclusion

THE ENGINEER AS STRUCTURAL ARTIST

"The real cycle you're working on is a cycle called yourself. The machine that appears to be 'out there' and the person that appears to be 'in here' are not two separate things. They grow toward Quality or fall away from Quality together."

—ROBERT M. PIRSIG, *ZEN AND THE ART OF MOTORCYCLE MAINTENANCE*[1]

Engineers are the professionals to whom our society has given the responsibility of designing bridges. Our society asks that its bridges perform well, cost no more than necessary, and look good. The public wants structures that are assets to their communities, in every sense of that word.

Those facts impose upon us, as engineers, the responsibility to learn how to make structures attractive. Other professionals have contributions to make, but it is our responsibility and cannot be delegated. Aesthetics must take its place alongside performance and economy as a criterion of bridge design. The achievement of aesthetic quality must take its place alongside performance and economy as skills of bridge engineers. Aesthetics is not more important than the other two, but neither should it be any less important.

Aesthetic awareness will give each individual engineer the potential to create structural art. Engineers such as John Roebling and Gustave Eiffel in earlier times, and Robert Maillart, Christian Menn, Jean Muller, Jörg Schlaich, and others of our own time, have given us inspiring examples of this art (Figure C-1). Any engineer with any bridge can aspire to these standards. While he or she may not always attain them, the experience and the product will be better for the effort.

Unfortunately, the achievement of structural art cannot be guaranteed simply from the application of these

FIGURE C-1 *Structural art at Glade Creek in West Virginia. The pier responds to the forces on it while providing a massive contrast to the spidery truss.*

guidelines. The guidelines can help the designer avoid the worst errors, provide a standard of basic quality, and point him or her in the right direction. But the achievement of great art is beyond prescription; it depends on a unique confluence of the problem, the resources, and the sensibility of the designer.

This book aims to help develop their sensibility; but, in the end, the result will depend on judgments and ideas the designer must find within—judgments and ideas honed by looking at, thinking about, and designing the best possible bridges.

---

1   Pirsig, Robert M., 1974. *Zen and the Art of Motorcycle Maintenance,* New York: William Morrow and Company.

ILLUSTRATION CREDITS

CHAPTER ONE

Fig. 1-1 Photograph courtesy of J. Wayman Williams.

Fig. 1-2 Photo: © Baron Wolman (copyright).

Fig. 1-3 *Bridge Aesthetics Around the World*. (1991 by Transportation Research Board, National Research Council. Reprinted by permission.)

Fig. 1-4 Eric Hyne, Illustrator, based on photographs by Fritz Leonhardt.

Fig. 1-5 Eric Hyne, Illustrator, based on photographs by Fritz Leonhardt.

Fig. 1-6 Eric Hyne, Illustrator, based on photographs by Fritz Leonhardt.

Fig. 1-9.2 *Bridge Aesthetics Around the World*. (1991 by Transportation Research Board, National Research Council. Reprinted by permission.)

Fig. 1-10.2 Photograph courtesy of Portland Cement Association, Skokie, IL.

Fig. 1-11.1 *Bridge Aesthetics Around the World*. (1991 by Transportation Research Board, National Research Council. Reprinted by permission.)

Fig. 1-11.2 Photograph courtesy of Vic Anderson, Delcan Corporation, Toronto, Canada.

Fig. 1-13 Photograph courtesy of National Steel Bridge Alliance and Texas Department of Transportation.

Fig. 1-14 Billington, David, 1979. *Robert Maillart's Bridges, The Art of Engineering*. Princeton, NJ: Princeton University Press. Reprinted by permission.

Fig. 1-15 Billington, David, 1979. *Robert Maillart's Bridges: The Art of Engineering*. Princeton, NJ: Princeton University Press. Reprinted by permission.

Fig. 1-16 Photograph courtesy of Mancia/Bodmer, FBM Studio.

Fig. 1-17 Billington, David, 1990. *Robert Maillart and the Art of Reinforced Concrete*. Architectural History Foundation and Massachusetts Institute of Technology. Reprinted by permission of MIT Press.

Fig. 1-18 Photograph courtesy of Mancia/Bodmer, FBM Studio.

Fig. 1-19 Photograph courtesy of Mancia/Bodmer, FBM Studio.

Fig. 1-20 Photograph courtesy of Jon Wallsgrove.

Fig. 1-21 Benevolo, Leonardo, 1971. *History of Modern Architecture*. Cambridge, MA: MIT Press. Reprinted by permission.

Fig. 1-22 Photograph courtesy of J. Wayman Williams.

Fig. 1-23.1 Photograph courtesy of J. Wayman Williams.

Fig. 1-23.2 Photograph courtesy of J. Wayman Williams.

Fig. 1-25 Arenal Bridge, Cordoba. Designer/Photographer: Julio Martinez-Calzon.

Fig. 1-26 Arenal Bridge, Cordoba. Designer/Photographer: Julio Martinez-Calzon.

Fig. 1-27 Photograph courtesy of Juan Jose Arenas de Pablo.

CHAPTER TWO

Fig. 2-1.1 Photograph courtesy of Portland Cement Association, Skokie, IL.

Fig. 2-3 Photograph courtesy of Portland Cement Association, Skokie, IL.

Fig. 2-6 Eric Hyne, illustrator, based on sketches by the author.

Fig. 2-7 Photograph courtesy of Elk River Concrete Products.

Fig. 2-8.1 Photograph courtesy of John Minor.

Fig. 2-9 Photograph courtesy of Portland Cement Association, Skokie, IL.

Fig. 2-10 Photo courtesy of URS Corp.

Fig. 2-11 Photo courtesy of Maryland State Highway Administration

Fig. 2-17 Drawing by Alicia Buchwalter, based on sketches by the author.

Fig. 2-18 Eric Hyne, illustrator, based on sketches by the author.

Fig. 2-20 Photograph courtesy of Hed Grouni and Ontario Ministry of Transportation.

Fig. 2-23 Drawing by Alicia Buchwalter, based on sketches by the author.

Fig. 2-25 Drawing by Alicia Buchwalter, based on sketches by the author.

Fig. 2-26 Photograph courtesy of Portland Cement Association, Skokie, IL.

Fig. 2-27 Photograph courtesy of Maryland State Highway Administration, Baltimore, MD. Maryland Route 213 over the Sassafras River. Designer: Envirodyne Engineers, Inc.

Fig. 2-28 Photograph courtesy of Maryland State Highway Administration, Baltimore, MD. Maryland Route 213 over the Sassafras River. Designer: Envirodyne Engineers, Inc.

Fig. 2-30 Photograph courtesy of David Billington.

Fig. 2-31 Photograph courtesy of Fritz Leonhardt.

Fig. 2-32 Photographer: Michael Freeman. Courtesy of Fruitlands Museums, Harvard, MA.

Fig. 2-33 Photo courtesy of the Portland Cement Association, IMG number 13878

Fig. 2-34 Drawing by Alicia Buchwalter, based on sketches by the author.

Fig. 2-36 Route 1369 over the Watauga River, Washington County, TN. Photograph courtesy of Tennessee Department of Transportation, Nashville, TN. Designer and Photographer: George Hornal.

Fig. 2-37 Made available for the National Steel Bridge Alliance/American Institute of Steel Construction promotional purposes.

Fig. 2-38 Photograph courtesy of Portland Cement Association, IMG number 14540.

Fig. 2-40 Robert DeLaunay, *Red Eiffel Tower (Le Tour rouge)*, 1911–1912. Photograph courtesy of Solomon R. Guggenheim Foundation, New York.

Fig. 2-41 Photograph courtesy of Christian Menn.

Fig. 2-43 Photograph courtesy of Emily Webster.

Fig. 2-44 Photograph courtesy of Christian Menn.

Fig. 2-45.1 Photograph courtesy of Christian Menn.

Fig. 2-45.2 Brancusi, Constantin (1876–1957) © ARS, NY. *Bird in Space*. 1928. Bronze (unique cast) 54 x 8 1/2 x 6 1/2". Given anonymously. (153-1934) The Museum of Modern Art, New York, NY. © 2004 Artists Rights Society (ARS), New York, ADAGP, Paris. Digital Image © The Museum of Modern Art/ Licensed by SCALA/ Art Resource, NY.

Fig. 2-46 Photograph courtesy of Fritz Leonhardt.

Fig. 2-48 Photograph courtesy of Steve Ladish.

Fig. 2-50 Photograph courtesy of Portland Cement Association, Skokie, IL.

Fig. 2-51 Photograph courtesy of Portland Cement Association, Skokie, IL.

Fig. 2-53 Melrose Interchange, Nashville, TN. Photograph courtesy of Tennessee Department of Transportation, Nashville, TN. Designer and Photographer: George Hornal.

Fig. 2-55 Sleater-Kinney Road Bridge, Interstate 5 Olympia, WA. Washington State DOT Bridge and Structures Office, Olympia, WA. Designer and Photographer: R. Ralph Mays, Architect, Olympia, WA.

Fig. 2-56 Photograph courtesy of Portland Cement Association, Skokie, IL.

Fig. 2-57 Photo courtesy of R.V. Arnaudo.

CHAPTER THREE

Fig. 3-2 Photograph courtesy of Fritz Leonhardt.

Fig. 3-3 Made available for the National Steel Bridge Alliance/American Institute of Steel Construction promotional purposes.

Fig. 3-4 Photograph courtesy of Portland Cement Association, Skokie, IL.

Fig. 3-5 Photograph courtesy of Portland Cement Association, IMG number 14586.

Fig. 3-6 Photograph courtesy of J. Wayman Williams.

Fig. 3-7.1 *Bridge Aesthetics Around the World*. (1991 by Transportation Research Board, National Research Council. Reprinted by permission.)

Fig. 3-7.2 Made available for the National Steel Bridge Alliance/ American Institute of Steel Construction promotional purposes.

Fig. 3-8 Photograph courtesy of Michael Baker Jr., Inc.

Fig. 3-11 Photograph courtesy of Michael Baker Jr., Inc.

Fig. 3-12.1 Eric Hyne, illustrator, based on sketches by the author.

Fig. 3-12.2 Eric Hyne, illustrator, based on sketches by the author.

Fig. 3-15.1 Photograph courtesy of Maryland State Highway Administration, Baltimore, MD. US Route 50 over US Route 301. Designer: Maryland State Highway Administration. Photographer: Marvin Blimline.

Fig. 3-15.2 Photograph courtesy of Maryland State Highway Administration, Baltimore, MD. US Route 50 over US Route 301. Designer: Maryland State Highway Administration. Photographer: Marvin Blimline.

Fig. 3-18 Photo courtesy of the Portland Cement Association, IMG number 68671.

Fig. 3-20 Photo courtesy of the Portland Cement Association, IMG number 13798.

Fig. 3-22 Photograph courtesy of Conrad Bridges.

Fig. 3-23 Drawings courtesy of HDR Engineering, Folsom, CA.

Fig. 3-26 Photo courtesy of HDR Engineering and City of Clearwater, FL.

Fig. 3-28 Photograph courtesy of CH2M Hill for the Iowa Department of Transportation.

Fig. 3-29 Photograph courtesy of CH2M Hill for the Iowa Department of Transportation.

Fig. 3-30.1 Photograph courtesy of CH2M Hill for the Iowa Department of Transportation.

Fig. 3-30.2 Photograph courtesy of CH2M Hill for the Iowa Department of Transportation.

Fig. 3-30.3 Photograph courtesy of CH2M Hill for the Iowa Department of Transportation.

Fig. 3-30.4 Photograph courtesy of CH2M Hill for the Iowa Department of Transportation.

CHAPTER FOUR

Fig. 4-4 Made available for the National Steel Bridge Alliance/American Institute of Steel Construction promotional purposes.

Fig. 4-5 Drawing courtesy of Michael Baker, Jr., Inc.

Fig. 4-6 Rendering by Rosales Gottemoeller & Associates, Inc.

Fig. 4-8 Photograph courtesy of Chicago Department of Transportation.

Fig. 4-9 Photograph courtesy of Portland Cement Association, Skokie, IL.

Fig. 4-11 Made available for the National Steel Bridge Alliance/American Institute of Steel Construction promotional purposes.

Fig. 4-12 Drawing by Alicia Buchwalter, based on sketches by the author.

Fig. 4-13 Photograph courtesy of Beiswenger Hoch & Associates.

Fig. 4-14 Made available for the National Steel Bridge Alliance/American Institute of Steel Construction promotional purposes.

Fig. 4-15 Drawing by Alicia Buchwalter, based on sketches by the author.

Fig. 4-16 Eric Hyne, illustrator, based on sketches by the author.

Fig. 4-17 Photograph courtesy of Portland Cement Association, IMG number 13454.

Fig. 4-18.1 Eric Hyne, illustrator, based on sketches by the author.

Fig. 4-18.2 Eric Hyne, illustrator, based on sketches by the author.

Fig. 4-20 Eric Hyne, illustrator, based on sketches by the author.

Fig. 4-21 Photograph courtesy of Maryland State Highway Administration, Baltimore, MD. Blooming Rose Road over I-68. Designer: Ewell, Baumhardt & Associates. Photographer: Marvin Blimline.

Fig. 4-22 Drawing by Alicia Buchwalter, based on sketches by the author.

Fig. 4-23.1 Eric Hyne, illustrator, based on sketches by the author.

Fig. 4-23.2 Eric Hyne, illustrator, based on sketches by the author.

Fig. 4-24.1 Eric Hyne, illustrator, based on sketches by the author.

Fig. 4-24.2 Eric Hyne, illustrator, based on sketches by the author.

Fig. 4-25 Eric Hyne, illustrator, based on sketches by the author.

Fig. 4-26 Eric Hyne, illustrator, based on sketches by the author.

Fig. 4-27.1 Drawing by Alicia Buchwalter, based on sketches by the author.

Fig. 4-27.2 Drawing courtesy of J. Wayman Williams.

Fig. 4-28 Photograph courtesy of Hed Grouni and Ontario Ministry of Transportation.

Fig. 4-29 Photograph courtesy of David Billington.

Fig. 4-30 Eric Hyne, illustrator, based on sketches by the author.

Fig. 4-31 Eric Hyne, illustrator, based on sketches by the author.

Fig. 4-32 Eric Hyne, illustrator, based on sketches by the author.

Fig. 4-33 Eric Hyne, illustrator, based on sketches by the author.

Fig. 4-34 Eric Hyne, illustrator, based on sketches by the author.

Fig. 4-35.1 Eric Hyne, illustrator, based on sketches by the author.

Fig. 4-35.2 Eric Hyne, illustrator, based on sketches by the author.

Fig. 4-35.3 Eric Hyne, illustrator, based on sketches by the author.

Fig. 4-36 Eric Hyne, illustrator, based on sketches by the author.

Fig. 4-37 Eric Hyne, illustrator, based on sketches by the author.

Fig. 4-39 Photograph courtesy of Stewart Watson.

Fig. 4-40.1 Eric Hyne, illustrator, based on sketches by the author.

Fig. 4-40.2 Eric Hyne, illustrator, based on sketches by the author.

Fig. 4-43 Photographs courtesy of North Carolina Department of Transportation, Raleigh, NC.

Fig. 4-44 Photograph courtesy of Richard Serra. Photographer: Dirk Reinartz.

Fig. 4-45 Eric Hyne, illustrator, based on sketches by the author.

Fig. 4-46 *Bridge Aesthetics Around the World.* (1991 by Transportation Research Board, National Research Council. Reprinted by permission.)

Fig. 4-47 Eric Hyne, illustrator, based on sketches by the author.

Fig. 4-48 Eric Hyne, illustrator, based on sketches by the author.

Fig. 4-49 Photograph courtesy of Maryland State Highway Administration, Baltimore, MD. Old Montgomery Road over I-95. Designer: Green Associates. Photographer: M.A. Zulkowski.

Fig. 4-50 Eric Hyne, illustrator, based on sketches by the author.

Fig. 4-51 Eric Hyne, illustrator, based on sketches by the author.

Fig. 4-52 Eric Hyne, illustrator, based on sketches by the author.

Fig. 4-55 Eric Hyne, illustrator, based on sketches by the author.

Fig. 4-56 Eric Hyne, illustrator, based on sketches by the author.

Fig. 4-57 Photograph courtesy of Portland Cement Association, Skokie, IL.

Fig. 4-58 Eric Hyne, illustrator, based on sketches by the author.

Fig. 4-59 Eric Hyne, illustrator, based on sketches by the author.

Fig. 4-60.1 Eric Hyne, illustrator, based on sketches by the author.

Fig. 4-60.2 Eric Hyne, illustrator, based on sketches by the author.

Fig. 4-61 Eric Hyne, illustrator, based on sketches by the author.

Fig. 4-63 Drawing by Alicia Buchwalter, based on sketches by the author.

Fig. 4-65 Eric Hyne, illustrator, based on sketches by the author.

CHAPTER FIVE

Fig. 5-1 Drawing by Alicia Buchwalter, based on sketches by the author.

Fig. 5-2 Eric Hyne, illustrator, based on sketches by the author.

Fig. 5-3 Illustration by Gabriel Ross, based on sketches by the author.

Fig. 5-4 Eric Hyne, illustrator, based on sketches by the author.

Fig. 5-5 Illustration by Gabriel Ross, based on sketches by the author.

Fig. 5-7 Eric Hyne, illustrator, based on sketches by the author.

Fig. 5-8 Eric Hyne, illustrator, based on sketches by the author.

Fig. 5-9 Photograph courtesy of Stewart Watson.

Fig. 5-10 Eric Hyne, illustrator, based on sketches by the author.

Fig. 5-11 Drawing by Alicia Buchwalter, based on sketches by the author.

Fig. 5-12 Eric Hyne, illustrator, based on sketches by the author.

Fig. 5-13 Eric Hyne, illustrator, based on sketches by the author.

Fig. 5-14 Drawing by Alicia Buchwalter, based on sketches by the author.

Fig. 5-15 Eric Hyne, illustrator, based on sketches by the author.

Fig. 5-16 Eric Hyne, illustrator, based on sketches by the author.

Fig. 5-17 Eric Hyne, illustrator, based on sketches by the author.

Fig. 5-18 Eric Hyne, illustrator, based on sketches by the author.

Fig. 5-19 Eric Hyne, illustrator, based on sketches by the author.

Fig. 5-20.1 Photograph courtesy of California Department of Transportation, Sacramento, CA.

Fig. 5-21 Photograph by Neil Kyeberg, Minnesota Department of Transportation.

Fig. 5-25 *Bridge Aesthetics Around the World.* (1991 by Transportation Research Board, National Research Council. Reprinted by permission.)

Fig. 5-26.1 Eric Hyne, illustrator, based on sketches by the author.

Fig. 5-26.2 Eric Hyne, illustrator, based on sketches by the author.

Fig. 5-26.3 Eric Hyne, illustrator, based on sketches by the author.

Fig. 5-27 Eric Hyne, illustrator, based on sketches by the author.

Fig. 5-28 Photo courtesy of the Portland Cement Association, IMG number 13797.

Fig. 5-29.1 Eric Hyne, illustrator, based on sketches by the author.

Fig. 5-29.2 Eric Hyne, illustrator, based on sketches by the author.

Fig. 5-29.3 Eric Hyne, illustrator, based on sketches by the author.

Fig. 5-30 Photograph courtesy of Kinzelman Kline, Columbus, Ohio.

Fig. 5-32 Eric Hyne, illustrator, based on sketches by the author.

Fig. 5-33 Eric Hyne, illustrator, based on sketches by the author.

Fig. 5-34 Eric Hyne, illustrator, based on sketches by the author.

Fig. 5-35.1 Eric Hyne, illustrator, based on sketches by the author.

Fig. 5-37.1 Photograph courtesy of Maryland State Highway Administration, Baltimore, MD. Looking North (Orange), Looking South (Green). Designer: Envirodyne Engineers, Inc. Photographer: Marvin Blimline.

Fig. 5-37.2 Photograph courtesy of Maryland State Highway Administration, Baltimore, MD. Looking North (Orange), Looking South (Green). Designer: Envirodyne Engineers, Inc. Photographer: Marvin Blimline.

Fig. 5-38 "Controlling Color" (Patricia Lambert, 1991 Design Press).

Fig. 5-39 Linda Holtzschue, 2002. *Understanding Color.* New York: John Wiley & Sons, Inc. This material is used by permission of John Wiley & Sons, Inc.

Fig. 5-40 "Controlling Color" (Patricia Lambert, 1991 Design Press).

Fig. 5-41 Linda Holtzschue, 2002. *Understanding Color.* New York: John Wiley & Sons, Inc. This material is used by permission of John Wiley & Sons, Inc.

Fig. 5-43 Photograph courtesy of Stewart Watson.

Fig. 5-44 Made available by the National Steel Bridge Alliance/American Institute of Steel Construction promotional purposes.

Fig. 5-46 Harrison Pike over I-81, Kingsport, TN. Photograph courtesy of Tennessee Department of Transportation, Nashville, TN. Designer and Photographer: George Hornal.

Fig. 5-56 Photo courtesy of the Portland Cement Association, IMG number 13800.

Fig. 5-57 Photograph courtesy of Stewart Watson.

Fig. 5-59 Sleater-Kinney Road Bridge, Interstate 5 Olympia, WA. Washington State DOT Bridge and Structures Office, Olympia, WA. Designer and Photographer: R. Ralph Mays, Architect, Olympia, WA.

Fig. 5-62.1 Eric Hyne, illustrator, based on sketches by the author.

Fig. 5-62.2 Eric Hyne, illustrator, based on sketches by the author.

Fig. 5-64 Eric Hyne, illustrator, based on sketches by the author.

Fig. 5-66 Photo courtesy of the Portland Cement Association, IMG number 13799.

Fig. 5-67 Photographs courtesy of North Carolina Department of Transportation, Raleigh, NC.

Fig. 5-68 Eric Hyne, illustrator, based on sketches by the author.

Fig. 5-69 Photograph Courtesy of Kinzelman Kline.

Fig. 5-70 City of Phoenix, Arizona, Office of Arts and Culture; Jody Pinto, Artist. Photograph by Craig Smith.

Fig. 5-71 Photograph courtesy of Stewart Watson.

Fig. 5-72 Photograph courtesy of Maryland State Highway Administration, Baltimore, MD. Maryland Route 175 EB over Patapsco River. Designer: Green Associates.

CHAPTER SIX

Fig. 6-2 Photograph courtesy of J. Wayman Williams.

Fig. 6-3 Granby Road over I-81, Kingsport, TN. Photograph courtesy of Tennessee Department of Transportation,

Nashville, TN. Designer and Photographer: George Hornal.

Fig. 6-4 Photograph courtesy of Portland Cement Association, Skokie, IL.

Fig. 6-10.1 Photograph courtesy of David Billington.

Fig. 6-10.2 Photograph courtesy of David Billington.

Fig. 6-10.3 Photograph courtesy of David Billington.

Fig. 6-11 Photograph courtesy of J. Wayman Williams.

Fig. 6-15.1 Photograph courtesy of J. Wayman Williams.

Fig. 6-15.2 Photograph courtesy of Portland Cement Association, Skokie, IL.

Fig. 6-21 Photograph courtesy of J. Wayman Williams.

Fig. 6-22 Photo courtesy of Indiana Department of Transportation.

Fig. 6-23 Made available by the National Steel Bridge Alliance for promotional purposes.

Fig. 6-26.1 State Route 62 over White Creek, Morgan County, TN. Photograph courtesy of Tennessee Department of Transportation, Nashville, TN. Designer and Photographer: George Hornal.

Fig. 6-27.1 Photograph courtesy of Portland Cement Association, Skokie, IL.

Fig. 6-27.2 Photograph courtesy of Portland Cement Association, IMG number 14589.

Fig. 6-28 Photograph courtesy of Portland Cement Association, IMG number 14368.

Fig. 6-29.1 Photograph courtesy of Portland Cement Association, IMG number 14589.

Fig. 6-29.2 Photograph courtesy of Portland Cement Association, IMG number 14585.

Fig. 6-30.1 Photo courtesy of the Portland Cement Association, IMG number 13802.

Fig. 6-30.2 Photo courtesy of the Portland Cement Association, IMG number 13801.

Fig. 6-31 State Route 20 over the Tennessee River, Perry County, TN. Photograph courtesy of Tennessee Department of Transportation, Nashville, TN. Designer and Photographer: George Hornal.

Fig. 6-32 *Bridge Aesthetics Around the World*. (1991 by Transportation Research Board, National Research Council. Reprinted by permission.)

Fig. 6-33 Photograph courtesy of Portland Cement Association, Skokie, IL.

Fig. 6-35 Photograph courtesy of Portland Cement Association, Skokie, IL.

Fig. 6-36 Photograph courtesy of Portland Cement Association, Skokie, IL.

Fig. 6-37 Photograph courtesy of Portland Cement Association, Skokie, IL.

Fig. 6-38 State Route 20 over the Tennessee River, Perry County, TN. Photograph courtesy of Tennessee Department of Transportation, Nashville, TN. Designer and Photographer: George Hornal.

Fig. 6-40 Photograph courtesy of Portland Cement Association, Skokie, IL.

Fig. 6-41 Photograph courtesy of Portland Cement Association, Skokie, IL.

Fig. 6-44 Photograph courtesy of HDR Engineering and City of Clearwater, FL.

Fig. 6-45 Photograph courtesy of HDR Engineering and City of Clearwater, FL.

Fig. 6-46 Photograph courtesy of Fritz Leonhardt.

Fig. 6-47 Photograph courtesy of Portland Cement Association, Skokie, IL.

Fig. 6-48 Photograph courtesy of Portland Cement Association, Skokie, IL.

Fig. 6-49 Photograph courtesy of Portland Cement Association, IMG number 14587.

Fig. 6-50 Photograph courtesy of David Billington.

Fig. 6-51 Made available for the National Steel Bridge Alliance/American Institute of Steel Construction promotional purposes.

Fig. 6-52 State Route 386 over I-65, 2-Mile Pike and CSX RR. Photograph courtesy of Tennessee Department of Transportation, Nashville, TN. Designer and Photographer: George Hornal.

Fig. 6-53 Photograph courtesy of Portland Cement Association, Skokie, IL.

Fig. 6-54 Photograph courtesy of Portland Cement Association, IMG number 14584.

Fig. 6-60 Photograph courtesy of URS Corp. Photographer: Jim Pillips.

Fig. 6-61 Photograph courtesy of Hardesty and Hanover.

Fig. 6-62 Photograph courtesy of E.C. Driver & Associates, Consulting Engineers. 17th Street Causeway Bridge. Engineer-of-Record James M. Phillips III, PE.

Fig. 6-63 Photograph courtesy of National Steel Bridge Alliance and Texas Department of Transportation.

Fig. 6-64 Photograph courtesy of Hardesty and Hanover.

Fig. 6-65 Weems Creek Bridge, URS, Thomas D. Jenkins, photo by Jeffrey G. Katz.

Fig. 6-66 Photograph courtesy of URS Corp. 17th Street Causeway Bridge. Photographer: David Lawrence. Engineer-of-Record James M. Phillips III, PE. E.C. Driver & Associates Consulting Engineers.

Fig. 6-67.1 Drawing courtesy of URS Corp. Architect: Todd Rose, Helman Hurley Charvat Peacock Architects.

Fig. 6-67.2 Drawing courtesy of URS Corp. Architect: Todd Rose, Helman Hurley Charvat Peacock Architects.

Fig. 6-68 Photograph courtesy of URS Corp. 17th Street Causeway Bridge. Photographer: David Lawrence. Engineer-of-Record James M. Phillips III, PE., E.C. Driver & Associates Consulting Engineers.

Fig. 6-69 Photograph courtesy of URS Corp. 17th Street Causeway Bridge. Photographer: David Lawrence. Engineer-of-Record James M. Phillips III, P. E., E.C. Driver & Associates Consulting Engineers.

Fig. 6-70 Photograph courtesy of URS Corp. 17th Street Causeway Bridge. Photographer: David Lawrence. Engineer-of-Record James M. Phillips III, PE., E.C. Driver & Associates Consulting Engineers.

Fig. 6-71 Photograph courtesy of URS Corp. 17th Street Causeway Bridge. Photographer: David Lawrence. Engineer-of-Record James M. Phillips III, PE. E.C. Driver & Associates Consulting Engineers.

Fig. 6-72 Made available for the National Steel Bridge Alliance/American Institute of Steel Construction promotional purposes.

Fig. 6-73 Photograph courtesy of Portland Cement Association, Skokie, IL.

Fig. 6-74.2 Eric Hyne, illustrator, based on sketches by the author.

Fig. 6-75 Photograph courtesy of Maryland State Highway Administration, Baltimore, MD. Maryland Route 175 EB over Patapsco River. Designer: Green Associates.

Fig. 6-76 Made available for the National Steel Bridge Alliance/American Institute of Steel Construction promotional purposes.

Fig. 6-77 Photograph courtesy of Jörg Schlaich.

Fig. 6-78.1 Drawing courtesy of Jörg Schlaich.

Fig. 6-78.2 Drawing courtesy of Jörg Schlaich.

Fig. 6-79.1 Photograph courtesy of Jörg Schlaich.

Fig. 6-79.2 Photograph courtesy of Jörg Schlaich.

Fig. 6-80 Photograph courtesy of Jörg Schlaich.

Fig. 6-81 Photograph courtesy of California Department of Transportation, Sacramento, CA.

Fig. 6-84 Photograph courtesy of Portland Cement Association, Skokie, IL. IMG number 14588.

Fig. 6-85 Photograph courtesy of Portland Cement Association, Skokie, IL.

Fig. 6-86 Photograph courtesy of Portland Cement Association, Skokie, IL. IMG number 13878.

CHAPTER SEVEN

Fig. 7-2 Photograph courtesy of J. Wayman Williams.

Fig. 7-3 Photograph courtesy of Portland Cement Association, Skokie, IL.

Fig. 7-5 Photograph courtesy of David Billington.

Fig. 7-6 Photograph courtesy of Maryland State Highway Administration.

Fig. 7-8 Photograph courtesy of Yvonne Thelwell, Lawrie & Associates Consulting Engineers.

Fig. 7-9.1 Image prepared by URS Corp; courtesy of Maryland State Highway Administration.

Fig. 7-9.2 Image prepared by URS Corp; courtesy of Maryland State Highway Administration.

Fig. 7-9.3 Image prepared by URS Corp; courtesy of Maryland State Highway Administration.

Fig. 7-10 Image prepared by URS Corp; courtesy of Maryland State Highway Administration.

Fig. 7-11 Image prepared by URS Corp; courtesy of Maryland State Highway Administration.

Fig. 7-12 Image prepared by URS Corp; courtesy of Maryland State Highway Administration.

CONCLUSION

Fig. C-1 Photograph courtesy of Steve Ladish.

INDEX